A-Level Maths for OCR

M1

Paul Sanders

Text © Nelson Thornes 2005
Original illustrations © Nelson Thornes Ltd 2005

The right of Paul Sanders to be identified as author of this work has been asserted by him in accordance with the Copyright, Designs and Patents Act 1988.

Published in 2005 by:
Nelson Thornes Ltd
Delta Place
27 Bath Road
CHELTENHAM
GL53 7TH
United Kingdom

05 06 07 08 09 / 10 9 8 7 6 5 4 3 2 1

A catalogue record for this book is available from the British Library

ISBN 0 7487 9456 5

Sample paper written by David Lee

Page make-up by Mathematical Composition Setters Ltd, Salisbury, United Kingdom

Printed and bound in Spain by Graphycems

Acknowledgements

We are grateful to the Oxford Cambridge and RSA Examination Board for permission to reproduce all the questions marked OCR.
All answers provided for examination questions are the sole responsibility of the author.

The publishers have made every effort to contact copyright holders but apologise if any have been overlooked.

CONTENTS

A-Level Maths for OCR is a brand new series from Nelson Thornes designed to give you the best chance of success in Advanced Level Maths. This book fully covers the OCR **M1** module specification.

In each chapter, you will find a number of key features:

- A beginning of chapter **OBJECTIVES** section, so you can see clearly what you should learn from each chapter

- **WORKED EXAMPLES** taking you through common questions, step by step

- Carefully graded **EXERCISES** to give you thorough practice in all concepts and skills

- Highlighted **KEY POINTS** to help you see at a glance what you need to know for the exam

- An **IT ICON** (IT) to highlight areas where IT software such as Excel may be used

- **EXTENSION** boxes with background information and additional theory

- An end-of-chapter **SUMMARY** to help with your revision

- An end-of-chapter **REVISION EXERCISE** so you can test your understanding of the chapter

At the end of the book, you will find a **MODULE REVISION EXERCISE** containing exam-type questions for the entire module. This is divided into five sections, mirroring the structure of the specification. [Each section tells you which chapters you should have done.]

Finally, there is a **SAMPLE EXAM PAPER** written by an OCR examiner which you can do under timed exam conditions to see just how well prepared you are for the real exam.

1 Introduction to Mathematical Modelling

The purpose of this chapter is to enable you to

- understand the concept of a mathematical model

Mathematical Modelling

Mechanics is the study of how and why objects move or don't move. As a mathematical subject, the techniques and skills of pure maths, together with the observational laws of physics, are brought together so that predictions about the motion of bodies can be made.

The process of applying mathematical techniques to real life problems is sometimes called "Mathematical Modelling". Mathematical equations produce a "model" which, it is hoped, will be a realistic representation of the real situation.

The process of creating a mathematical model can be usefully summarised by the "modelling cycle":

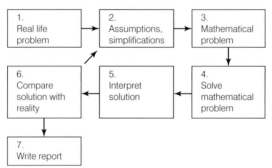

Starting with the real life problem, which needs to be clearly stated and understood, thought must be given to how the problem might be solved. In many cases, the problem may be so complicated that progress can only be made by making appropriate assumptions to simplify the problem.

The simplified problem should then be expressible as a mathematical problem whose solution can be found using standard mathematical techniques.

Having obtained a solution to the mathematical problem, it is essential to interpret the answer in the context of the original real life problem and use one's experience of the real life problem to decide whether the solution is satisfactory or not. If the solution is satisfactory then a final account explaining the solution can be prepared, but if the model does not give realistic information about the real life problem then the assumptions and simplifications made early in the modelling process must be reconsidered and another loop of the modelling cycle must be undertaken.

This modelling cycle will be illustrated using a financial example in which the cheapest means of buying a car is to be found.

Step 1: State the real life problem

John is buying a new car and is choosing between the following payment schemes:

Scheme A:	Cash purchase of £15 000
Scheme B:	Short-term loan:
	Deposit of £7500 and 24 monthly payments of £375, the first monthly payment to be made one month after the purchase of the car.
Scheme C:	Long-term loan:
	Deposit of £3750 and 36 monthly payments of £360, the first monthly payment to be made one month after the purchase of the car.

Which purchase scheme is the most advantageous to John?

Step 2: Simplifications and assumptions

Assume that John has sufficient capital available to choose any of these three methods.

Ignore what John does with any unused money.

Step 3: The mathematical problem

Which of the three schemes has the lowest overall cost?

Step 4: Solve the mathematical problem

The costs of each scheme can be calculated as:

Scheme A   £15 000

Scheme B   7500 + 24 × 375 = £16 500

Scheme C   3750 + 36 × 360 = £16 710

Step 5: Interpret the solution

The cash scheme is the best solution, being £1500 cheaper than the short-term loan and £1710 cheaper than the long-term loan.

Step 6: Compare with reality

The way the overall cost of the loan has been calculated does not take any account of the interest that John could earn from a building society or bank account if he didn't spend all the money at once – this interest may be sufficient to change the best solution.

The modelling cycle must be looped through again trying to take interest on savings into account.

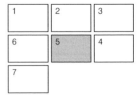

Step 2': Simplifications and assumptions

Assume that John has sufficient capital available to choose any of these three methods.

Assume that John invests any of the £15 000 capital that he hasn't already spent in a building society paying 4% per annum interest.

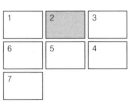

Step 3': The mathematical problem

What are the overall costs of each of the three schemes?

Step 4': Solve the mathematical problem

The cost of scheme A remains at £15 000.

To calculate the costs of schemes B and C we need to know the monthly interest rate equivalent to 4% per annum.

Suppose $p\%$ per month is equivalent to 4% per annum then

$$\left(1 + \frac{p}{100}\right)^{12} = 1.04$$

$$\Rightarrow \quad 1 + \frac{p}{100} = 1.04^{1/12} = 1.0032737 \ldots$$

$$\Rightarrow \quad p = 0.32737 \ldots .$$

Consider now scheme B:

If x_n denotes the amount of the initial £15 000 left in the building society just before the nth monthly repayment then the sequence of values x_n, satisfies

$$x_1 = 7500 \times 1.0032737 \ldots = 7524.55$$
$$x_2 = (x_1 - 375) \times 1.0032737 \ldots = 7172.96$$
$$x_3 = (x_2 - 375) \times 1.0032737 \ldots = 6820.21$$

At the end of month 1 there is £x_1 in the account. At the beginning of month 2 a repayment of £375 is made which leaves £$(x_1 - 375)$ in the account which grows to £$(x_1 - 375) \times 1.0032737 \ldots$ by the end of the month.

and, in general, x_{n+1} is linked to x_n by the equation

$$x_{n+1} = (x_n - 375) \times 1.0032737 \ldots .$$

The values of x_n can be calculated using either a spreadsheet or a graphical calculator:

Month	Savings at end of month
1	7524.55
2	7172.96
3	6820.21
4	6466.31
5	6111.25
6	5755.03
7	5397.64
8	5039.08
9	4679.35
10	4318.44
11	3956.35
12	3593.07
13	3228.61
14	2862.95
15	2496.09
16	2128.03
17	1758.77
18	1388.30
19	1016.62
20	643.72
21	269.60
22	
23	
24	

From the table we can see that the account has not sufficient funds to pay the 21st and subsequent instalments.

The extra money required is
21st instalment = 105.40
22nd instalment = 375.00
23rd instalment = 375.00
24th instalment = 375.00
making a total of £1230.40.

The total cost to John of this purchase scheme is therefore £16 230.40

The third scheme can be analysed in a similar fashion.

If y_n denotes the amount of the initial £15 000 left in the building society just before the nth monthly repayment, then y_n satisfies the recurrence relation

$$y_1 = 11\ 250 \times 1.0032737 \ldots = 11\ 286.83$$
$$y_{n+1} = (y_n - 360) \times 1.0032737 \ldots$$

and the spreadsheet or calculator shows that

$$y_{34} = 7.42$$

so this scheme requires the following cash inputs:
34th instalment = 352.58
35th instalment = 360.00
36th instalment = 360.00
making a total of £1072.58.

The total cost to John of this scheme is therefore £16 072.58

Step 5': Interpret the solution

The cash scheme is still the best solution, being £1230.40 cheaper than the short-term loan and £1072.58 cheaper than the long-term loan.

John should therefore pay cash if there are no other considerations to be taken into account. If, however, he wishes to make use of one of the loan options, scheme C looks the better option.

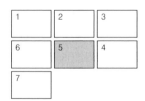

Step 6': Compare with reality

This time, each of the three schemes has been considered in a way which takes into account the interest that could be earned on capital that has not been paid at the time of purchase.

An extra loop around the modelling cycle might attempt to take into account the possibility of changes in interest rates.

Step 7: Write a report

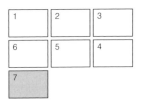

Once a satisfactory solution to the problem has been found, a report should be prepared to summarise and communicate the major findings of the model making process. The nature of this report will vary from problem to problem and must also be appropriate to the mathematical ability of the potential reader.

In some cases it may be appropriate to simply state the conclusions that can be drawn from the modelling process, but frequently it is necessary to give details of the working that led to the conclusions.

For the example being considered, if John has a non-mathematical friend then the final report to him might be

"Scheme A gives the cheapest way of buying the car. The two loan schemes will end up costing over £1000 more than the cash purchase scheme even when interest on savings is taken into account.

If other factors make a loan scheme desirable, then the long-term loan works out slightly cheaper than the short-term loan when interest on savings is taken into account."

EXERCISE 1

Suggestion: Working in pairs or small groups, attempt just one or two of the questions in this exercise. Try to use the modelling cycle to develop a mathematical model for the real life problem.

1 A TV sports programme is planning a "Battle of The Sexes" athletics event in which each competitor – male or female – will run three races over three distances:

     300 m   500 m   2000 m

The athlete is to be awarded a number of points for his/her performance in each event.

The organisers wish to devise a scoring system for these races which will give fair handicaps to the two sexes and give equal weighting to each of the races.

The table below shows the world record times (as at May 2005) for the established athletics events:

Event	100 m	200 m	400 m	800 m	1500 m	5000 m
Male record(s)	9.78	19.32	43.18	101.11	206.00	757.35
Female record(s)	10.49	21.34	47.60	113.28	230.46	864.68

Use this information to devise a fair points scheme for this "Battle of The Sexes" event.

2 The "Supercomp" computer company has regional offices in London, Exeter, Manchester and Edinburgh and the table on the right shows the number of people employed in each office.

Office	Number of employees
London	400
Exeter	100
Manchester	300
Edinburgh	300

The company is proposing to build a new headquarters and it is expected that people from each of the regional offices will have to regularly visit this new headquarters.

Advise the company about where to site the office.

3 Speed humps are to be placed on a street passing a primary school to try and ensure that vehicles do not travel faster than 50 km/hr along the street. If a car goes over the humps at more than 25 km/hr then the passengers in the car will receive a nasty jolt.

How far apart should the speed humps be placed in order to keep the speed of cars down to safe levels?

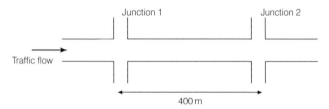

d metres

4 A busy one-way street in a town has two junctions with lesser roads and these junctions are 400 m apart.

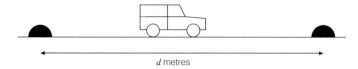

Traffic lights are to be placed at both junctions. Suggest the size of the time lapse there should be between the traffic lights on the main street at junction 1 changing to green and the lights on the main street at junction 2 changing to green.

Modelling the Motion of Objects

The mathematical models that we will develop to describe the motion of objects all rely on three laws formalised by Sir Isaac Newton in the seventeenth century. Phrased in modern English, these three laws are:

Newton's First Law
Every body will remain at rest or continue to move with constant velocity unless there is a non-zero resultant force acting on it.

Newton's Second Law
When an external resultant force is applied to a body, there is an acceleration whose magnitude is proportional to the resultant force and which is in the same direction as the resultant force.

Newton's Third Law
If a first body exerts a force on a second body then the second body exerts a force of equal size but in the opposite direction on the first body.

Although these laws are results based on experiments and observations rather than being mathematical theorems that can be rigorously proved, their value comes from the fact that they lead to mathematical models that accurately predict the motion of most objects.

The model only fails when consideration is given to the motion of very small or very large objects and when consideration is given to bodies moving at speeds approaching the speed of light: in these cases the techniques of Quantum Mechanics or Relativity are required.

Newton's three laws relate the **forces** acting on a body to the motion of the body as measured by the **velocity** and **acceleration** of the object. Most people have intuitive ideas about what is meant by "force", "velocity" and "acceleration" : in chapters 2 and 4 these ideas will be formalised so that Newton's laws can be applied in chapters 3 and 5 to situations involving the motion of objects.

Common Modelling Assumptions in Mechanics

Step 2 in the modelling cycle involved making assumptions and simplifications to the real life problem to produce a revision of the initial problem that can be analysed satisfactorily. In initial studies of Mechanics the following are examples of simplifications that are frequently made:

- objects are treated as particles whose size is negligible
- some surfaces are treated as being perfectly smooth so there is no friction when an object moves across the surface
- the resistance of the air to the motion of a body is assumed to be negligible
- strings connecting two objects are assumed to be both light and inextensible
- collisions between objects may be assumed to be instantaneous

These, and other, assumptions will be discussed in more detail in later chapters.

Having studied this chapter you should

- understand the process of mathematical modelling and, in particular, realise that assumptions and simplifications are often required in producing a model and that these may lead to results that do not accurately reflect real life

2 Forces

The purpose of this chapter is to enable you to

- draw accurate force diagrams for objects
- find the component of a force in a given direction
- find the resultant of a system of forces

Newton's three laws of motion tell us that an understanding of the forces acting on an object is critical if predictions are to be made about how the object is going to move. In this chapter, we move from an informal intuition about forces to the point where we can accurately portray all the forces acting on an object. Then, using trigonometry, we see how the forces can be simplified analytically to determine the resultant force acting on the object.

Representing Forces

If a man is pushing a crate across a floor then we say that he is exerting a force on the crate.
The pushing force on the crate can be shown in a diagram by an arrow acting in the direction of the force.
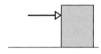

As a convention, throughout this book, forces will be shown as arrows with a non-filled triangular arrowhead. (Different types of arrows will be used later to signify other quantities possessing both size and direction.)

If the man now ties a rope around the crate and pulls the crate across the floor then the crate is being pulled to the left by the force in the rope which is usually called the **tension**.
This tension can easily be shown in a diagram.
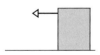

If the man now lifts the crate off the ground and holds it still then the crate will experience a supporting force from the man's arms.

This force can also be shown diagrammatically.

Examples of Forces in Mechanics

The force that we are most aware of in our daily lives is the gravitational attraction force exerted by the Earth on every object on or close to the Earth's surface. This force is usually called the **weight** of the object and acts through the centre of mass of the object.

> **Modelling Assumption**
> We shall assume that the weight of an object is the same at all points on or near the Earth's surface. In fact the weight of an object varies slightly from point to point on the Earth's surface and decreases as the object moves away from the Earth's surface.

If a body is being pulled by a string, spring or chain then the body is being pulled by the **tension** in the string, spring or chain. Similarly, if a body is being pushed by a compressed spring or a rod then it is being pushed by the **thrust** in the spring or rod.

The engine of a vehicle produces a **driving force**.

If a body is in contact with another body then there is always a **contact** force between the two bodies. It is usual to split this contact force into two parts:

> **a normal reaction** or **normal force** which is always present and acts perpendicular to the line of contact of the two objects

and

> **a frictional force** which acts along the line of contact if the contact is not perfectly smooth and there is some motion (or tendency towards motion) between the two bodies.

A special case of a frictional force is **air resistance** which can be regarded as the frictional force between a moving object and the surrounding air.

> **Modelling Assumption**
> In many examples in this module, we will assume that the air resistance is negligible compared to the other forces. This assumption may be reasonable if the motion is taking place in a vacuum or if the object moving can be regarded as a particle or if the speed of the object is low.

When dealing with the motion of a vehicle it is quite common to combine the frictional forces and air resistance forces to produce a single **resistance force** that acts in the opposite direction to the motion.

The Effect of a Force

Consider now the effect of a single force acting on a small block which is resting on a sheet of ice.

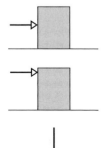

If the force applied to the block is horizontal and acts through the centre of the block then we would expect the block to accelerate in the direction of the force and we would also expect the size of the force to affect the size of the acceleration.

If the force applied to the block is horizontal and does not act through the centre of the block then we would expect the block to accelerate in the direction of the force and also to rotate.

If the force applied to the block is vertically downwards and acts through the centre of the block then we would not expect any motion unless the force was so large as to deform the block or crack the ice.

If the force is applied obliquely to the block then the block would be expected to slide and topple.

From these examples, we know that the effect of a force on a body depends on

- the size of the force
- the direction of the force
- the point of application of the force

Force is therefore a **vector** quantity: it has both size and direction. The S.I. unit of force is the Newton, which abbreviates to N. One Newton is approximately the weight of an apple. A formal definition of a Newton will be given in chapter 4.

S.I. (Systéme internationale) units should be used in Mechanics calculations. This international system of units is based on

- the metre as a unit of length
- the kilogram as a unit of mass
- the second as a unit of time
- the ampere as a unit of electrical current
- the degree Kelvin as a unit of temperature

All other S.I. units are defined in terms of these basic units. For example, the S.I. unit for speed is m/s or ms^{-1} and the S.I. unit for volume is m^3.

Force Diagrams

A key skill in Mechanics is the ability to draw a force diagram showing **all** the forces acting upon a particular body.

EXAMPLE 1

Show the forces acting on a block which is resting on a horizontal table.

The forces acting on the block are:

- the weight, W, of the block;
- a normal reaction, R, between the block and the table.

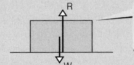

There is no frictional force since there is no possibility of the block moving in any direction.

EXAMPLE 2

Show the forces acting on a particle which is at rest on a rough inclined plane.

Modelling Assumption
A particle is an object whose size can be assumed to be negligible. Drawing a particle is impossible so in our diagrams a particle will be represented by a small rectangle or circle.

There will be three forces:

- the weight, W, of the particle;
- the normal reaction, R, between the particle and the inclined plane, acting perpendicular to the inclined plane;
- a frictional force, F, up the plane which opposes the possible motion of the object down the plane.

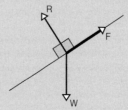

EXAMPLE 3

Show the forces acting on a car as it travels up a hill inclined at 8° to the horizontal.

> **Modelling Assumption**
> This will be shown as a single force but is really four separate forces acting at each of the points of contact between the car and the road: we show it as a single force since we are treating the car as a particle.

Treating the car as a particle, the forces are:

- the weight, W, of the car;
- the driving force, D, produced by the car's engine;
- the normal reaction, R, between the car and the road;
- a resistance force, P, which is the combination of the air resistance and frictional forces. This acts down the slope since the car is moving up the slope.

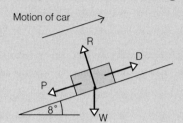

EXAMPLE 4

Show the forces acting on

a) a small stone after it has been dropped from the top of a tower 20 m high;
b) a free fall sky diver after she has jumped out of a plane flying at an elevation of 3000 m.

a) The only important force acting is the weight, W, of the stone.

> **Modelling Assumption**
> Since the stone can be regarded as a particle and the speeds are relatively low, air resistance will be ignored. If the conclusions drawn from the resulting model are unrealistic then an improved model, which takes air resistance into account, should be produced.

b) For the sky diver, the air resistance, P, is certainly not negligible, so the force diagram will be

EXERCISE 1

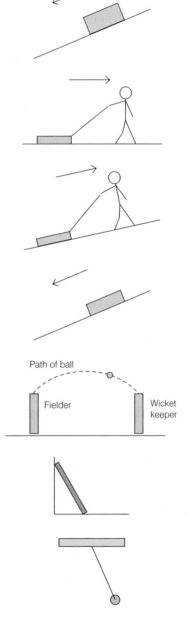

1 The diagram shows a particle sliding down a smooth plane. Copy the diagram and show all the forces acting on the particle.

2 The diagram shows a sledge being pulled across rough horizontal ground by a boy. Copy the diagram and show all the forces acting on the sledge.

3 The diagram shows a sledge being pulled up a rough hill by a boy. Copy the diagram and show all the forces acting on the sledge.

4 The diagram shows a sledge sliding down a rough hill. Copy the diagram and show all the forces acting on the sledge.

5 The diagram shows a cricket ball as it is in flight from a fielder to the wicket keeper. Copy the diagram and show all the forces acting on the ball at this instant.

6 The diagram shows a wooden plank resting against a rough horizontal wall and rough horizontal ground. Copy the diagram and show all the forces acting on the plank.

7 The diagram shows a pendulum consisting of a light string and a metal bob swinging in a vacuum. Copy the diagram and show the forces acting on the metal bob.

8 Draw a diagram to show the forces acting on a car that is being driven down a hill inclined at 5° to the horizontal.

9 Draw diagrams to show the forces which are acting on a small block which is
   a) at rest on a rough surface inclined at an angle of 20° to the horizontal;
   b) sliding down a smooth surface inclined at an angle of 30° to the horizontal;
   c) sliding down a rough surface inclined at an angle of 30° to the horizontal;
   d) pulled along a smooth horizontal surface by a string inclined at an angle of 20° to the horizontal;
   e) pulled up a rough slope inclined at an angle of 30° to the horizontal by a string inclined at an angle of 40° to the horizontal.

10 Draw a diagram showing the forces acting on a ladder which is leaning with one end against a smooth vertical wall and the other end standing on rough horizontal ground.

11 A space rocket is to be launched vertically upwards from a launch pad on the Earth's surface. Draw diagrams to show the forces acting on the rocket
   a) whilst it is at rest at the launch pad;
   b) when the rockets are firing and it has just left the launch pad;
   c) whilst it is in the Earth's atmosphere, travelling at 100 m/s with the rockets firing.

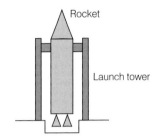

Force Diagrams for Systems of Objects

We will frequently need to draw diagrams showing either the forces acting on a complete system of objects or the forces acting on individual components of the system.

When drawing a force diagram for a whole system we **only** show the external forces acting on the system. The internal forces acting between parts of the system are ignored since Newton's third law tells us that these forces are equal in magnitude and opposite in direction so cancel each other out.

When drawing force diagrams for components of a system we do need to show the internal actions of one component on another. Newton's third law tells us that if component A exerts a force on component B then component B exerts an equal and opposite force on component A.

EXAMPLE 5

A car is pulling a trailer along a horizontal road. Draw separate force diagrams to show the forces acting on

a) the whole system of car and trailer;
b) the trailer only;
c) the car only.

EXAMPLE 5 (continued)

a) The forces acting on the **whole system** are:
- the weight, W_c, of the car and the weight, W_t, of the trailer;
- the driving force, D, produced by the car's engine;
- the normal reaction, R_c, between the car and the road and the normal reaction, R_t, between the trailer and the road;
- the total resistance, P_c, to the motion of the car and the total resistance, P_t, to the motion of the trailer.

The force diagram
for the whole system is

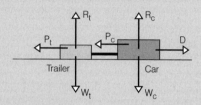

b) The forces acting on just the **trailer** are:
- the weight, W_t, of the trailer;
- the normal reaction, R_t, between the trailer and the road;
- the total resistance, P_t, to the motion of the trailer;
- a force, T, acting in the direction of motion, along the tow-bar pulling the trailer.

The diagram showing the
forces acting on the trailer is

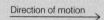

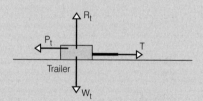

c) The forces acting on just the **car** are:
- the weight, W_c, of the car;
- the driving force, D, produced by the car's engine;
- the normal reaction, R_c, between the car and the road;
- the total resistance, P_c, to the motion of the car;
- a force, T, acting in the opposite direction to the motion, in the tow-bar linking the car to the trailer.

The diagram showing the
forces acting on the car is

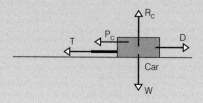

EXAMPLE 6

The diagram shows a structure made from three blocks of wood.

Draw separate force diagrams to show

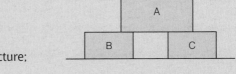

i) the forces acting on the whole structure;
ii) the forces acting on block A;
iii) the forces acting on block B.

i) The weights of the three blocks (W_A, W_B and W_C) and the normal reactions (R_B and R_C) between the ground and the two bottom blocks are the only external forces acting on the **whole structure**.

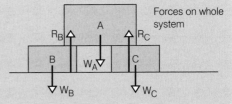

ii) The only forces acting on **block A** are its weight and two supporting forces, P_B and P_C, from blocks B and C.

iii) There are three forces acting on **block B**:
- its weight;
- the normal reaction between the ground and B;
- the force exerted by block A on block B. Newton's third law says that this is equal and opposite to the force exerted by block B on block A.

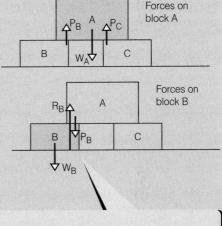

Important Note
The force acting down from A on B is **NOT** the weight of A but is the normal force due to the contact between A and B.
Remember that the weight of A is a force that acts on A and not on any other body.

EXERCISE 2

1 The diagram shows a tower made out of two rectangular blocks. The tower is standing on a horizontal floor.

a) Draw a force diagram to show the forces acting on the whole tower.
b) Draw a force diagram to show the forces acting on the top block.
c) Draw a force diagram to show the forces acting on the bottom block.

2 The diagram shows Saffron using a string to pull her toy train across the carpet. Her toy train consists on an engine and two trucks.

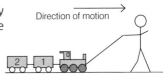

a) Draw a diagram to show the forces acting on the whole train.
b) Draw a diagram to show the forces acting on the engine.
c) Draw a diagram to show the forces acting on truck 1.
d) Draw a diagram to show the forces acting on truck 2.

3 A ladder rests in a vertical plane with one end against a rough vertical wall and the other end on rough horizontal ground. There is a block suspended from the ladder by a string attached to a point one-third of the way up the ladder.

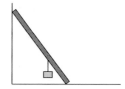

Draw separate diagrams to show the forces acting on
a) the whole system;
b) the block;
c) the ladder.

4 The diagram shows a simplified model of a lift with a passenger.

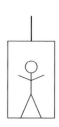

Draw separate diagrams to show the forces acting on
a) the passenger;
b) the lift.

Writing Forces as Vectors: Resolving Forces

Since forces are vector quantities, column vectors can be used to give an easily manipulated numerical form for forces.

Consider, for example, the force, F, in the diagram which has a magnitude of 8 N and acts in a direction which is inclined 32° above the positive x direction.

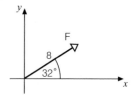

The force can be expressed as a vector by creating a right-angled triangle whose hypotenuse is the force and whose two other sides are parallel to the co-ordinates axes:

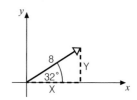

Trigonometry can be used to find the sizes of X and Y:

$$\frac{X}{8} = \cos 32° \Rightarrow X = 8\cos 32° \Rightarrow X = 6.784 \qquad \text{(3 d.p.)}$$

$$\frac{Y}{8} = \sin 32° \Rightarrow Y = 8\sin 32° \Rightarrow Y = 4.239 \qquad \text{(3 d.p.)}$$

so the force can be written as the **column vector** $\begin{pmatrix} 8\cos 32° \\ 8\sin 32° \end{pmatrix}$, which is $\begin{pmatrix} 6.784 \\ 4.239 \end{pmatrix}$ correct to three decimal places.

We say that, for horizontal and vertical axes x and y,

8 cos 32° is the x **component** or the **horizontal component** of the force F

and that

8 sin 32° is the y **component** or the **vertical component** of the force F.

EXAMPLE 7

A force P has magnitude 12 N and acts in a direction that makes an angle with the positive x direction of 113° in an anticlockwise direction. Express P as a column vector.

$$\frac{X}{12} = \sin 23° \Rightarrow X = 12 \sin 23° \Rightarrow X = 4.689 \quad \text{(3 d.p.)}$$

$$\frac{Y}{12} = \cos 23° \Rightarrow Y = 12 \cos 23° \Rightarrow Y = 11.046 \quad \text{(3 d.p.)}$$

so

$$P = \begin{pmatrix} -4.689 \\ 11.046 \end{pmatrix} \quad \text{(3 d.p.)}$$

Notice that the horizontal part of P is acting to the left so the horizontal component of P must be a *negative* value.

The process of splitting a force into two components acting in perpendicular directions is called **resolving the force**. So far we have resolved forces into horizontal and vertical components, but a force can be resolved into any two mutually perpendicular directions.

In particular, when considering motion on an inclined plane it is usual to resolve the forces in directions that are parallel and perpendicular to the inclined plane.

EXAMPLE 8

A particle of weight W is pulled up a rough plane inclined at 17° to the horizontal by a string making an angle of 30° with the horizontal.

a) Draw a diagram to show all the forces acting on the particle.
b) Find the components of each of these forces in directions that are parallel and perpendicular to the inclined plane.

The forces acting on the particle are

- its weight, W;
- the normal reaction with the plane, R;
- the frictional force, F, acting down the plane and opposing the motion up the plane;
- the tension in the string, T.

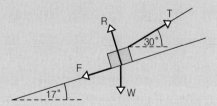

EXAMPLE 8 (continued)

The weight can be split into components parallel and perpendicular to the plane by drawing a right-angled triangle whose hypotenuse is the weight and whose other two sides are parallel and perpendicular to the plane.

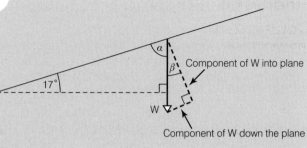

Component of W into plane

Component of W down the plane

Elementary triangle geometry implies that

$\alpha = 73°$ and $\beta = 17°$.

Trigonometry now gives

component of W into the plane = W cos 17°

and

component of W down the plane = W sin 17°.

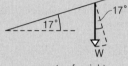

components of weight

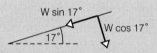

The normal reaction N acts perpendicular to the plane so we can immediately write

component of N out of the plane = N

and

component of N down the plane = 0.

The friction F acts down the plane so we can immediately write

component of F out of the plane = 0

and

component of F down the plane = F.

Finding the components of T parallel and perpendicular to the plane requires more care. Again a right-angled triangle must be drawn with T as the hypotenuse and with the other two sides parallel and perpendicular to the direction of the plane.

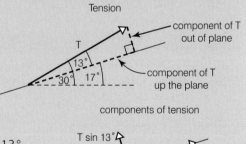

Tension

component of T out of plane

component of T up the plane

components of tension

From this diagram, we can see that

component of T out of the plane = T sin 13°

and

component of T up the plane = T cos 13°.

Finding the Magnitude and Direction of a Force from its Components

Pythagoras's theorem and trigonometry can also be used to find the magnitude and direction of a force from knowledge of its two components.

EXAMPLE 9

A force D has a horizontal component of 6.3 N and a vertical component of −3.8 N. Find the magnitude and direction of D.

Start by drawing a diagram to illustrate the force.

Let D be the magnitude of D.

Using Pythagoras's theorem:

$$D^2 = 6.3^2 + 3.8^2$$
$$\Rightarrow \quad D^2 = 54.13$$
$$\Rightarrow \quad D = \sqrt{54.13}$$
$$\Rightarrow \quad D = 7.357 \qquad \text{(3 d.p.)}$$

Since the vertical component of D is negative, D acts in a direction **below** the x-axis.

Suppose D makes an angle θ with the x-axis:

$$\tan \theta = \frac{3.8}{6.3} \Rightarrow \theta = 31.1° \qquad \text{(3 s.f.)}$$

So D has magnitude 7.357 N and acts in a direction of 31.1° below the positive x direction.

EXERCISE 3

1 Find the horizontal and vertical components of each of the following forces:

a)

b)

c)

d)

2 Express each of the following forces as column vectors:

a)
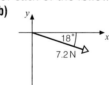

b)
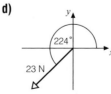

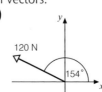

3 Calculate the magnitude of the horizontal and vertical components of
   a) a force of 6 N inclined at 20° to the horizontal;
   b) a force of 20 N inclined at 48° to the vertical;
   c) a force of 50 N inclined at 67° to the horizontal.

4 In each of the following diagrams a force is shown acting on a particle which is on an inclined plane. Find the components of this force in directions that are parallel and perpendicular to the inclined plane and make it clear whether the components are up or down the plane and in or out of the plane, respectively.

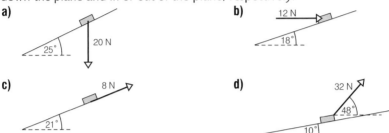

a)

b)

c)

d)

5 a) Resolve a weight of 25 N in two directions which are parallel and perpendicular to a plane inclined at 30° to the horizontal.
   b) Resolve a weight of 40 N in two directions which are parallel and perpendicular to a plane inclined at 12° to the horizontal.
   c) Resolve a weight of W in two directions which are parallel and perpendicular to a plane inclined at θ to the horizontal.

6 What are the components parallel to and perpendicular to an incline of 10° of a horizontal force of 30 N?

7 a) A force P has horizontal component 7.2 N and vertical component 8.9 N. Find the magnitude and direction of P.
   b) A force Q has horizontal component −37.2 N and vertical component 18.9 N. Find the magnitude and direction of Q.
   c) A force R is given by the vector $\begin{pmatrix} 12.3 \\ -8.9 \end{pmatrix}$. Find the magnitude and direction of R.

The Resultant of a System of Forces

The resultant of a system of forces is the single force that has the same effect as the system of forces.

Suppose that a sledge is being pulled by two forces over an icy pond, as shown in the diagram.

The resultant force can be found geometrically by drawing an accurate **parallelogram of forces**. The two forces are shown as two adjacent sides, OA and OC, of a parallelogram, OABC, and the resultant force is then given by the diagonal OB of the parallelogram.

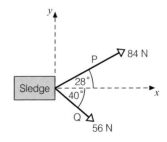

Using 1 mm in the diagram to represent a force of 1 N, the diagram would be

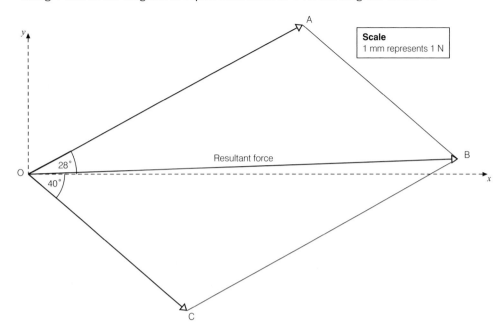

The line OB is 117 mm long and makes an angle of 2° with the *x* direction so the resultant force has a magnitude of 117 N and acts in a direction of 2° in an anticlockwise direction from the *x* direction.

Using a scale drawing to find the resultant force is unsatisfactory for two important reasons: it is time-consuming and tends to be inaccurate unless great care is taken in drawing the parallelogram.

The need for an accurate scale drawing can be removed entirely if we express the forces as column vectors.

x component of P = 84 cos 28°

y component of P = 84 sin 28°

so

$$P = \begin{pmatrix} 84 \cos 28° \\ 84 \sin 28° \end{pmatrix}$$

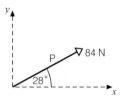

x component of Q = 56 cos 40°

y component of Q = −56 sin 40°

so

$$Q = \begin{pmatrix} 56 \cos 40° \\ -56 \sin 40° \end{pmatrix}$$

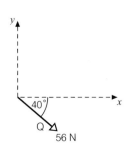

In the parallelogram of forces, the resultant force is represented by the vector $\overrightarrow{OB}$.

We know that the movement $\overrightarrow{OB}$ is the same as moving from O to A and then moving from A to B. In vector language, we write

$$\overrightarrow{OB} = \overrightarrow{OA} + \overrightarrow{AB}$$

Since OABC is a parallelogram,

$$\overrightarrow{AB} = \overrightarrow{OC}$$

so we can write

$$\overrightarrow{OB} = \overrightarrow{OA} + \overrightarrow{OC}$$

which means that

vector for resultant force = vector for P + vector for Q

$$\Rightarrow \quad \text{vector for resultant force} = \begin{pmatrix} 84 \cos 28° \\ 84 \sin 28° \end{pmatrix} + \begin{pmatrix} 56 \cos 40° \\ -56 \sin 40° \end{pmatrix}$$

$$= \begin{pmatrix} 84 \cos 28° + 56 \cos 40° \\ 84 \sin 28° - 56 \sin 40° \end{pmatrix}$$

$$= \begin{pmatrix} 117.066 \ldots \\ 3.4395 \ldots \end{pmatrix}$$

Now that the vector form of the resultant force is known, the techniques of the previous section can be used to find the magnitude and direction of the resultant force:

$$R^2 = 117.066\ldots^2 + 3.4395\ldots^2$$

$$\Rightarrow \quad R = 117.1 \text{ N} \quad (1 \text{ d.p.})$$

$$\tan\theta = \frac{3.4395\ldots}{117.066\ldots}$$

$$\Rightarrow \quad \theta = 1.7° \quad (1 \text{ d.p.})$$

So the two forces pulling the sledge are equivalent to a single force of magnitude 117.1 N acting in a direction of 1.7° in an anticlockwise direction from the x direction.

This method has two advantages over drawing an accurate parallelogram of forces; greater accuracy can be obtained and the method easily extends to cases when there are more than two forces pulling the sledge.

In general, if forces expressed as vectors $F_1, F_2, \ldots, F_n$ are applied to a particle then the vector form for the resultant force is given by $F_1 + F_2 + \cdots + F_n$.

EXAMPLE 10

Find the resultant force of the system of forces shown in the diagram, giving your answer as a magnitude and a direction.

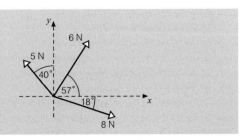

22

EXAMPLE 10 (continued)

SOLUTION 1

The vector form for the 6 N force is $\begin{pmatrix} 6\cos 57° \\ 6\sin 57° \end{pmatrix}$

The vector form for the 8 N force is $\begin{pmatrix} 8\cos 18° \\ -8\sin 18° \end{pmatrix}$

The vector form for the 5 N force is $\begin{pmatrix} -5\sin 40° \\ 5\cos 40° \end{pmatrix}$

> This is the **vector** form for the resultant force.

The resultant force is obtained by simply adding the three vectors:

$$\text{Resultant force} = \begin{pmatrix} 6\cos 57° \\ 6\sin 57° \end{pmatrix} + \begin{pmatrix} 8\cos 18° \\ -8\sin 18° \end{pmatrix} + \begin{pmatrix} -5\sin 40° \\ 5\cos 40° \end{pmatrix} = \begin{pmatrix} 7.6623\ldots \\ 6.3901\ldots \end{pmatrix}$$

Pythagoras's theorem and trigonometry can now be used to find the magnitude and direction of the resultant force:

$$R^2 = 7.6623\ldots^2 + 6.3901\ldots^2$$
$$\Rightarrow \quad R = 9.98 \text{ N} \quad (3 \text{ s.f.})$$
$$\tan\theta = \frac{6.3901}{7.6623}$$
$$\Rightarrow \quad \theta = 39.8° \quad (3 \text{ s.f.})$$

The resultant force has magnitude 9.98 N and makes an angle of 39.8° in an anticlockwise sense with the positive x direction.

The solution of this last example can be presented in a slightly different, and probably shorter, way by first calculating the total, X, of the horizontal components of each force and then calculating the total, Y, of the vertical components.

> This indicates we are resolving horizontally with "to the right" as the positive direction.

> The horizontal component of the 6 N force is 6 cos 57°.

> The horizontal component of the 8 N force is 8 cos 18°.

> The horizontal component of the 5 N force is −5 sin 40°.

SOLUTION 2

> The resultant force has a horizontal component of 7.6623...N and a vertical component of 6.3901...N

$$[\rightarrow] \quad X = 6\cos 57° + 8\cos 18° + (-5\sin 40°) = 7.6623\ldots$$
$$[\uparrow] \quad Y = 6\sin 57° + (-8\sin 18°) + 5\cos 40° = 6.3901\ldots$$

The magnitude and direction of the resultant force can now be calculated using basic trigonometry:

$$R^2 = 7.6623\ldots^2 + 6.3901\ldots^2$$
$$\Rightarrow \quad R = 9.98 \text{ N} \quad (3 \text{ s.f.})$$
$$\tan\theta = \frac{6.3901}{7.6623}$$
$$\Rightarrow \quad \theta = 39.8° \quad (3 \text{ s.f.})$$

The resultant force has magnitude 9.98 N and makes an angle of 39.8° in an anticlockwise sense with the positive x direction.

This shorter approach will be used in most examples but it should be remembered that a force is a vector and can therefore always be represented as a column vector.

EXAMPLE 11

The diagram shows a system of forces acting on a particle. Given that the resultant of this system of forces acts vertically, find

a) the magnitude of the force P;
b) the magnitude of the resultant force.

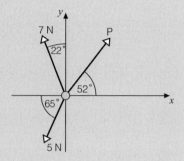

S
O
L
U
T
I
O
N

a) Resolving horizontally and vertically to find the components of the resultant force gives

$$[\rightarrow] \quad X = P \cos 52° + (-7 \sin 22°) + (-5 \cos 65°) = P \cos 52° - 4.735 \dots$$

$$[\uparrow] \quad Y = P \sin 52° + 7 \cos 22° + (-5 \sin 65°) = P \sin 52° + 1.958 \dots$$

We are told the resultant force is vertical so its horizontal component, X, must be zero:

$$P \cos 52° - 4.735 \dots = 0$$
$$\Rightarrow \quad P \cos 52° = 4.735 \dots$$
$$\Rightarrow \quad P = 7.691 \dots$$
$$\Rightarrow \quad P = 7.69 \quad \text{(3 s.f.)}$$

The magnitude of the force P is 7.69 N (3 s.f.)

b) The vertical component, Y, of the resultant force can now be calculated:

$$Y = P \sin 52° + 1.958 \dots = 8.0197 \dots$$

The resultant force is therefore a vertical force of magnitude 8.02 N (3 s.f.)

EXAMPLE 12

The diagram shows the forces acting on a particle placed on an inclined plane. Given that the net force acting perpendicular to the slope is zero, find

a) the value of R;
b) the magnitude and direction of the resultant force.

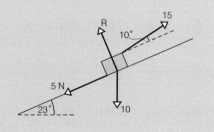

EXAMPLE 12 (continued)

a) We start by resolving each of the forces in directions parallel and perpendicular to the inclined plane:

	Component parallel to plane	Component perpendicular to plane
(10 N force, 23°)	10 sin 23° down the plane	10 cos 23° into the plane
(R force)	0	R out of the plane
(5 force)	5 down the plane	0
(15 force, 10°)	15 cos 10° up the plane	15 sin 10° out of the plane

The total of the components of the forces perpendicular to the plane is 0 so

$$R + 15 \sin 10° - 10 \cos 23° = 0$$
$$\Rightarrow \quad R = 6.60 \quad \text{(3 s.f.)}$$

> If "out of the plane" is treated as the positive direction, R and 15 have positive components perpendicular to the plane but the 10 N force has a negative component.

The force R is 6.60 N (3 s.f.)

b) Resolving parallel to the plane, taking up the plane as the positive direction,

total of the components of
the forces parallel to the plane = 15 cos 10° − 5 − 10 sin 23°
                                  = 5.864 ...

so the resultant force is 5.86 N (3 s.f.) acting up the plane.

EXERCISE 4

1 A sledge is pulled by the forces shown in the diagram. Find the magnitude and direction of the resultant force.

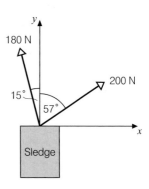

2 Find the total horizontal component and the total vertical component of each of the following force systems:

a)

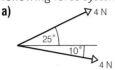

b)

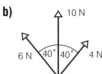

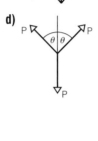

c)

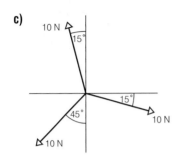

d)

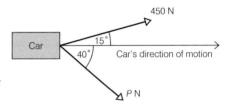

3 ABCD is a square. A force of 5 N acts along AB, a force of 8 N acts along AC and a force of 3 N acts along AD. Find the magnitude and direction of the resultant of these forces.

4 A car, which is initially at rest, is pulled by two boys with ropes as shown in the diagram:

Use the fact that the resultant force must be in the same direction as the car's motion to determine the value of P and the magnitude of the resultant force.

5 A block of weight 20 N is on a rough plane at 60° to the horizontal. The normal reaction is 10 N and the frictional force is 10 N acting up the plane. Find:
a) the total component parallel to the plane of these forces;
b) the total component perpendicular to the plane of these forces.

6 The diagram shows the forces acting on a block that is on an inclined plane.

If the resultant force of these forces is zero, find the values of R and F.

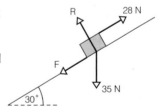

7 Three boys are pulling a sledge by means of three ropes. The boy in the middle is exerting a pull of 100 N. The other two boys, whose ropes each make an angle of 30° with the centre rope, are pulling with forces of 80 N and 140 N.
What is the resultant pull on the sledge and in which direction will it move?

8 ABCDEF is a regular hexagon.
Forces of magnitudes $2P$, $3P$, $4P$, $5P$ act along $\overrightarrow{AB}$, $\overrightarrow{AC}$, $\overrightarrow{AD}$, $\overrightarrow{AE}$, respectively.
Find the magnitude and direction of their resultant.

9 A force of magnitude $2\sqrt{2}$ N acts along the diagonal $\overrightarrow{AC}$ of a square ABCD. Another force of magnitude P acts along $\overrightarrow{AD}$. The resultant force is inclined at 60° to the side AB. Find the value of P and the magnitude of the resultant force.

10 Two forces P and Q have magnitudes 10 N and 12 N, respectively and act at the point O. The angle between P and Q is acute. The resultant of P and Q is R, and R makes an angle of 30° with Q, as shown in the diagram.

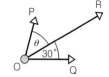

   a) By finding the total component of the forces P and Q in a direction perpendicular to the direction of R, show that $\theta = 36.87°$ (2 d.p.)
   b) Find the magnitude of R.

Having studied this chapter you should know how

- to draw force diagrams to show the forces acting on a particle or a simple object
- to express a force as a column vector, or equivalently, to resolve a force into its components in two perpendicular directions with particular emphasis on the cases of resolving horizontally and vertically and of resolving parallel and perpendicular to an inclined plane
- to find the magnitude and direction of a force from its vector form
- to find the resultant of a system of forces by expressing each of the forces as a vector and then adding the vectors

REVISION EXERCISE

1 The diagram shows two forces acting on a particle at O. The resultant force acts along the line Ox.
   a) Find the value of P.
   b) Find the magnitude of the resultant force.

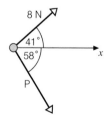

2 A force P makes an angle of 42° with the horizontal and has a horizontal component of 12.6 N. Find the magnitude of P. Find also the vertical component of P.

3 Two forces, each of magnitude 5 N, act at a point in the directions shown in the diagram. Find, in either order,
   i) the components in the x and y directions of the resultant of the two forces;
   ii) the magnitude and direction of the resultant.

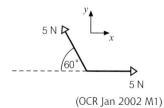

(OCR Jan 2002 M1)

4 The diagram shows a crane lifting a load, X, of weight 20 000 N. Attached to P is a cable, the other end of which is attached to a second load, Y, of weight 10 000 N.

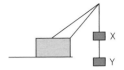

  i) Draw a diagram to show the forces acting on load Y.
  ii) Draw a diagram to show the forces acting on load X.

5 A force has horizontal and vertical components of 8 N and 12 N, respectively. Find the magnitude and direction of the force.

6 Forces of magnitude 8 N and 5 N act on a particle. The angle between the direction of the two forces is 30°. The resultant of the two forces has magnitude R N and acts at an angle $\theta°$ to the force of magnitude 8 N. Find R and θ.

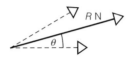

(OCR May 2002 M1)

7 Forces of magnitude 12 N and P N act on a particle in directions making an angle of 125°, as shown in the diagram. Given that the resultant acts in a direction which is perpendicular to the 12 N force, find P and the magnitude of the resultant.

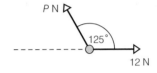

8 A sledge is pulled up a hill inclined at 10° to the horizontal by a rope inclined at 30° to the horizontal.

  i) Draw a diagram to show all the forces acting on the sledge.

  If the weight of the sledge is 200 N, the frictional force between the sledge and the hill is 80 N and the tension in the string is 140 N, calculate

  ii) the normal reaction between the sledge and the hill, given that the total component perpendicular to the hill of the forces is zero;
  iii) the total component parallel to the hill of the forces.

9 Forces P and Q act on a particle as shown in the diagram.

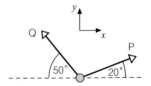

  The resultant of these forces has magnitude 16 N and makes an angle of 25° with the direction of P.
  i) By considering components of the forces in the x direction, prove that

$$P \cos 20° - Q \cos 50° = 16 \cos 45°$$

  ii) Write down a second equation that must be satisfied by P and Q.
  iii) Hence find the values of P and Q.

10 Find the magnitude and direction of the resultant of the system of forces shown in the diagram.

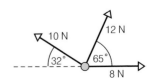

3 Equilibrium of Particles

The purpose of this chapter is to enable you to

- determine whether a particle is in equilibrium or not
- understand and use the rules for frictional forces

We have seen in chapter 2 how to calculate the resultant of a system of forces acting on a particle. Newton's first law states that

> **"If the resultant force is zero then the particle will either be at rest or be moving at constant speed in a straight line."**

We say that a particle is **in equilibrium** if the system of forces acting on a particle has a **zero resultant**. This means that a particle can only be in equilibrium if it is either at rest or moving in a straight line with constant speed.

Since the resultant of a system of forces can be found by finding the totals of the components of the forces in mutually perpendicular directions, the condition for a particle to be in equilibrium can be expressed in terms of the components of the forces.

In most cases that we will be considering the forces acting on the particle will be coplanar: that is, they are all acting in the same plane. The particle will then be in equilibrium if and only if **the totals of the components of all the forces in two mutually perpendicular directions in this plane are each zero**.

> You will meet these ideas in module C2.

Exact Values of Some Trigonometric Ratios

In order to prove that a particle is in equilibrium it must be shown that the resultant force is precisely zero. Almost zero is not good enough! Frequently a proof of equilibrium will require knowledge of the **exact** values of $\sin 30°$, $\cos 30°$, $\sin 45°$, $\cos 45°$, $\sin 60°$ and $\cos 60°$.

First, consider an equilateral triangle, ABC, whose sides are 2 units long. Let M be the foot of the perpendicular from C to the side AB.

Using Pythagoras's theorem

$$CM^2 = 2^2 - 1^2 = 3$$
$$\Rightarrow \quad CM = \sqrt{3}$$

Consideration of triangle AMC now gives

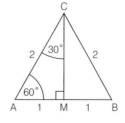

$$\sin 30° = \frac{AM}{AC} = \frac{1}{2}$$

$$\cos 30° = \frac{CM}{AC} = \frac{\sqrt{3}}{2}$$

$$\sin 60° = \frac{CM}{AC} = \frac{\sqrt{3}}{2}$$

$$\cos 60° = \frac{AM}{AC} = \frac{1}{2}$$

Now consider a right-angled isosceles triangle ABC with AB = BC = 1 unit.

Pythagoras's theorem gives

$$AC^2 = 1^2 + 1^2 = 2$$
$$\Rightarrow \quad AC = \sqrt{2}$$

so

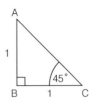

$$\sin 45° = \frac{AB}{AC} = \frac{1}{\sqrt{2}} = \frac{\sqrt{2}}{\sqrt{2}\sqrt{2}} = \frac{\sqrt{2}}{2}$$
$$\cos 45° = \frac{BC}{AC} = \frac{1}{\sqrt{2}} = \frac{\sqrt{2}}{\sqrt{2}\sqrt{2}} = \frac{\sqrt{2}}{2}$$

These results are summarised below. They need to be learnt!

$$\sin 30° = \frac{1}{2}, \qquad \cos 30° = \frac{\sqrt{3}}{2}$$
$$\sin 45° = \frac{\sqrt{2}}{2}, \qquad \cos 45° = \frac{\sqrt{2}}{2}$$
$$\sin 60° = \frac{\sqrt{3}}{2}, \qquad \cos 60° = \frac{1}{2}$$

Examples of Particle Equilibrium

EXAMPLE 1

A particle is subject to the forces shown in the diagram. Prove that the particle is in equilibrium.

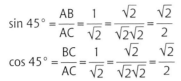

Let X denote the total of the horizontal components of the forces and Y denote the total of the vertical components. We must show that X = 0 and Y = 0.

$$X = 3 - 3\sqrt{2} \sin 45°$$
$$= 3 - 3\sqrt{2} \times \frac{\sqrt{2}}{2}$$
$$= 3 - \frac{3\sqrt{2} \times \sqrt{2}}{2}$$
$$= 3 - \frac{6}{2} = 0$$

$$Y = 4 - 1 - 3\sqrt{2} \cos 45°$$
$$= 3 - 3\sqrt{2} \cos 45°$$
$$= 3 - \frac{3\sqrt{2} \times \sqrt{2}}{2}$$
$$= 3 - \frac{6}{2} = 0$$

We have shown that the total components of these coplanar forces in two mutually perpendicular directions are both zero, so the particle is in equilibrium.

EXAMPLE 2

ABCDEF is a regular hexagon with centre O. A particle is held in equilibrium at O by forces of magnitudes 5, 2, 7, 4, P and Q acting in directions OA, OB, OC, OD, OE and OF, respectively.

Find P and Q.

> The usual labelling convention is that ABCDEF would be consecutive vertices of the hexagon if you went for a walk around the hexagon.

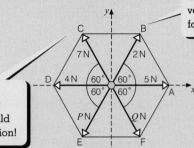

> Most Mechanics examples require a diagram: if a clear diagram is not given, it should be the first task in your solution!

Since the particle is in equilibrium under these forces

total of the components of the forces in the x direction = 0

$\Rightarrow \quad 5 + 2 \cos 60° - 7 \cos 60° - 4 - P \cos 60° + Q \cos 60° = 0$

$\Rightarrow \quad 5 + 1 - 3.5 - 4 - 0.5P + 0.5Q = 0$

$\Rightarrow \quad 0.5Q - 0.5P = 1.5$

$\Rightarrow \quad Q - P = 3$

> Remember: $\cos 60° = \frac{1}{2} = 0.5$.

and

total of the components of the forces in the y direction = 0

$\Rightarrow \quad 2 \sin 60° + 7 \sin 60° - P \sin 60° - Q \sin 60° = 0$

$\Rightarrow \quad 9 \sin 60° - P \sin 60° - Q \sin 60° = 0$

$\Rightarrow \quad 9 - P - Q = 0$

$\Rightarrow \quad Q + P = 9$

[÷ sin 60°]

The simultaneous equations $\left. \begin{array}{l} Q - P = 3 \\ Q + P = 9 \end{array} \right\}$ have solution $Q = 6$, $P = 3$.

> These equations can be solved using normal simultaneous equation methods or using the simultaneous equation solver on a graphical calculator.

EXAMPLE 3

One end of a string is attached to a fixed point A, the other is fastened to a small object of weight 12 N. The object is pulled aside by a horizontal force until the string is inclined at 20° to the vertical through A.

Find, correct to three significant figures, the magnitudes of the tension in the string and the horizontal force.

EXAMPLE 3 (continued)

Suppose the tension in the string is T N and the horizontal force is P N.

> Again, a clear force diagram of the situation is the first priority.

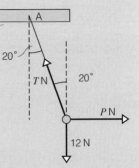

Since the particle is in equilibrium,

the total of the vertical components of the forces = 0

and

the total of the horizontal components of the forces = 0.

[↑] $T \cos 20° - 12 = 0$

$\Rightarrow \quad T \cos 20° = 12$

$\Rightarrow \quad T = \dfrac{12}{\cos 20°} = 12.770 \dots$

$\Rightarrow \quad T = 12.8 \qquad$ (3 s.f.)

> We will use the shorthand
>
> [↑]
>
> to signify the process of resolving vertically with up as the positive direction. Similarly, the shorthand
>
> [→]
>
> will be used to signify the process of resolving horizontally with "to the right" as the positive direction.

The tension in the string is 12.8 N (3 s.f.)

[→] $P - T \sin 20° = 0$

$\Rightarrow \quad P = T \sin 20° = 12.770 \dots \times \sin 20° = 4.367 \dots$

$\Rightarrow \quad P = 4.37$ N (3 s.f.)

The horizontal force is 4.37 N (3 s.f.)

EXAMPLE 4

A small block of weight 30 N is attached at C to two strings AC and BC of length 0.6 m and 0.8 m, respectively.

The ends A and B are attached to two points on a horizontal beam that are 1 m apart. Prove that the triangle ABC is right-angled and hence find the tension in each string when the block is hanging at rest below the beam.

A simple diagram is drawn to show the geometry of the problem:

Since

$$1^2 = 0.6^2 + 0.8^2$$

the converse of Pythagoras's theorem tells us that the triangle ABC is right-angled.

Now that we know the triangle ABC is right-angled we can write down some information about the angles in the triangles which will be useful when we start resolving:

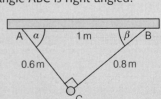

$\sin \alpha = \dfrac{0.8}{1} = 0.8 \qquad \cos \alpha = \dfrac{0.6}{1} = 0.6$

$\sin \beta = \dfrac{0.6}{1} = 0.6 \qquad \cos \beta = \dfrac{0.8}{1} = 0.8$

EXAMPLE 4 (continued)

A force diagram for the problem must now be drawn.

If T_1 and T_2 denote the tensions in the strings AC and BC, respectively then the force diagram for the forces acting on the particle is

Since the particle is in equilibrium,

the total of the vertical components of the forces = 0

and

the total of the horizontal components of the forces = 0.

> Remember that when a line intersects with two parallel lines then alternate angles are equal.

$$[\uparrow] \qquad T_1 \sin \alpha + T_2 \sin \beta - 30 = 0$$
$$\Rightarrow \quad 0.8T_1 + 0.6T_2 - 30 = 0$$
$$\Rightarrow \quad 0.8T_1 + 0.6T_2 = 30$$

$$[\rightarrow] \qquad -T_1 \cos \alpha + T_2 \cos \beta = 0$$
$$\Rightarrow \quad -0.6T_1 + 0.8T_2 = 0$$

> Substituting for the values of $\sin \alpha$ and $\sin \beta$ which were found earlier.

The simultaneous equations $\left. \begin{array}{l} 0.8T_1 + 0.6T_2 = 30 \\ -0.6T_1 + 0.8T_2 = 0 \end{array} \right\}$ have solution $T_1 = 24, \qquad T_2 = 18.$

The tension in the string CA is 24 N and the tension in the string CB is 18 N.

EXAMPLE 5

A particle of weight W is pulled up a smooth plane inclined at 30° to the horizontal at constant speed by a string inclined at 30° to the plane.

Find, in terms of W, the exact values of the normal reaction and the tension in the string.

Again, a clear force diagram for the particle is essential:

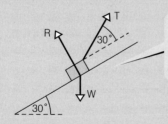

> Since the plane is smooth there is no friction. We are assuming that the air resistance on the particle is negligible.

The particle is moving at constant speed in a straight line so it is in equilibrium. This means that

the total of the components of the forces parallel to the plane = 0

and

the total of the components of the forces perpendicular to the plane = 0.

EXAMPLE 5 (continued)

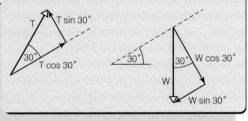

T has a component of T cos 30° up the plane;
W has a component of W sin 30° down the plane.

[parallel, up] $T\cos 30° - W\sin 30° = 0$

$$\Rightarrow \quad \frac{\sqrt{3}}{2}T - \frac{1}{2}W = 0$$

$$\Rightarrow \quad \frac{\sqrt{3}}{2}T = \frac{1}{2}W$$

$$\Rightarrow \quad T = \frac{W}{\sqrt{3}} = \frac{\sqrt{3}W}{\sqrt{3}\sqrt{3}}$$

$$\Rightarrow \quad T = \frac{\sqrt{3}W}{3}$$ [1]

We will use the shorthand
  [parallel, up]
to signify the process of resolving parallel to an inclined plane
with up as the positive direction.
Similarly, the shorthand
  [perpendicular, out]
will be used to signify the process of resolving perpendicular
to an inclined plane with "out of the plane" as the positive
direction.

[perpendicular, out] $R + T\sin 30° - W\cos 30° = 0$

$$\Rightarrow \quad R + \frac{1}{2}T - \frac{\sqrt{3}}{2}W = 0$$

$$\Rightarrow \quad R = \frac{\sqrt{3}}{2}W - \frac{1}{2}T$$

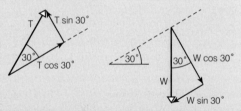

T has a component of T sin 30° out of the plane
W has a component of W cos 30° into the plane

Using [1] to substitute for T,

$$\Rightarrow \quad R = \frac{\sqrt{3}}{2}W - \frac{1}{2}\frac{\sqrt{3}W}{3} = \frac{\sqrt{3}}{2}W\left(1 - \frac{1}{3}\right) = \frac{\sqrt{3}}{2}W \times \frac{2}{3}$$

$$\Rightarrow \quad R = \frac{\sqrt{3}W}{3}$$

The normal reaction is $\frac{\sqrt{3}}{3}W$ and the tension in the string is $\frac{\sqrt{3}}{3}W$.

EXERCISE 1

1 Determine which of the following systems of forces are in equilibrium:

a)

2 N

45° 45°

1 N   1 N

b)

1 N

30°

√3 N

2 N

c)

5 N

5 N   5 N

60° 60°

2 ABCD is a square. A particle at A is subjected to three forces:
A force of 5 N acts along AB. A force of magnitude P acts along AD and a force of magnitude Q acts in direction CA. Given that the three forces are in equilibrium, find the magnitudes P and Q.

3 The diagram shows a particle which is moving at constant speed in a straight line under the action of three forces of magnitudes 6 N, 8 N and P N, respectively.

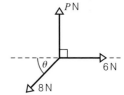

P N

θ

6 N

8 N

a) Find the angle θ and the value of P.

The force of magnitude P N is now removed.

b) State, or determine, the magnitude and direction of the resultant of the remaining two forces.

4 One end of a string is attached to a point A, and the other is fastened to a small object of weight 10 N. The object is pulled aside by a horizontal force until the string is inclined at 60° to the vertical through A. Find the tension in the string and the horizontal force.

5 One end of a light string of length 100 cm is attached to a point A on a vertical pole and the other end is fastened to a small object of weight 25 N. The object is pulled aside by a horizontal force until it is 35 cm away from the pole. Find the tension in the string and the horizontal force.

6 A small smooth ring, C, of weight 2 N is threaded onto a string whose ends are fixed to two points A and B in a horizontal line. The ring is pulled aside by a horizontal force of P N parallel to AB. When the ring is in equilibrium the section AC of the string is inclined at 40° to the vertical and BC is inclined at 20° to the vertical. Find the tension in the string and the value of P.

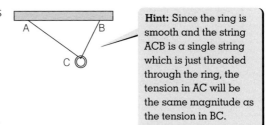

Hint: Since the ring is smooth and the string ACB is a single string which is just threaded through the ring, the tension in AC will be the same magnitude as the tension in BC.

7 A small object of weight 10 N rests in equilibrium on a rough inclined plane which makes an angle of 30° with the horizontal. Calculate the magnitude of the frictional force and the normal reaction.

8 A car of weight 12 000 N is driving at constant speed up a hill inclined at 8° to the horizontal. The total resistance (friction and air resistance) experienced by the car is 1800 N. Find the magnitude of the driving force produced by the car's engine.

9 The diagram shows the forces acting on a small block which is in equilibrium on a rough plane inclined at 25° to the horizontal.

Determine the magnitudes of the normal reaction, R, and the frictional force, F.

10 A weight of 26 N is supported by two strings AB and AC of lengths 0.5 m and 1.2 m, respectively. If BC is horizontal and of length 1.3 m:
a) prove that ∠BAC is a right angle;
b) calculate the magnitude of the tension in each of the two strings.

11 A small block of weight W rests on a smooth plane of inclination θ to the horizontal. Find the value of θ if

a) a force of $\frac{1}{2}$ W parallel to the plane is required to keep the block in equilibrium;

b) a horizontal force of $\frac{1}{3}$ W keeps the block in equilibrium.

Equilibrium of Connected Systems of Particles

The principles of the previous section can also be applied to the equilibrium of a system of particles. For equilibrium, the resultant force on the whole system must be zero and the resultant force on any component of the system must also be zero.

EXAMPLE 6

A car of weight 15 000 N pulls a caravan of weight 6000 N up a road which is inclined at 4° to the horizontal. The total resistance to the car's motion is 1300 N and the total resistance to the caravan's motion is 1200 N. If the car is moving with constant speed find

a) the driving force produced by the car's engine;
b) the force in the tow-bar connecting the caravan to the car.

State two assumptions that have been made in answering the question.

a) First draw a diagram to show the external forces acting on the **whole system** of car and caravan.

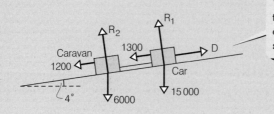

We do **not** show the force in the tow-bar in this diagram since it is an internal force of the whole system of car plus caravan.

EXAMPLE 6 (continued)

Assuming that the car is moving at constant speed in a straight line, the system will be in equilibrium so the resultant force acting on the system must be zero. In particular, the total of the components parallel to the slope of all of the forces must be zero:

$$[\text{parallel, up}] \quad D - 1300 - 1200 - 15\,000 \sin 4° - 6000 \sin 4° = 0$$
$$\Rightarrow \quad D = 3964.88\ ...$$
$$\Rightarrow \quad D = 3960 \qquad (3\ \text{s.f.})$$

The driving force produced by the car's engine is 3960 N (3 s.f.)

b) To calculate the force in the tow-bar, the forces on one part of the system (either the caravan or the car) must be considered. The diagram shows the forces acting on the caravan with P being the force in the tow-bar which pulls the caravan along.

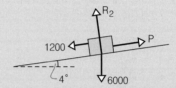

If the caravan is moving at constant speed in a straight line then it is in equilibrium so the total of the components parallel to the slope of all of the forces must be zero:

$$[\text{parallel, up}] \quad P - 1200 - 6000 \sin 4° = 0$$
$$\Rightarrow \quad P = 1618.53\ ...$$
$$\Rightarrow \quad P = 1620 \qquad (3\ \text{s.f.})$$

The force in the tow-bar connecting the caravan to the car is 1620 N (3 s.f.)

The assumptions are that

- the car and caravan are moving in a straight line;
- the car and caravan can be treated as if they were particles.

EXAMPLE 7

The diagram shows three small particles, P, Q and R, whose weights are 50 N, 40 N and 30 N, respectively. Particles P and Q are attached to the two ends of a string which passes over a small smooth fixed peg, A. Particles Q and R are attached to the ends of another string which passes round a second small smooth fixed peg, B. Particle R lies on a rough horizontal table.

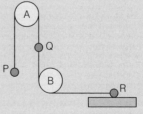

Given that the system is at rest, calculate

a) the tension in the string linking P and Q;
b) the tension in the string linking Q and R;
c) the frictional force acting on R.

EXAMPLE 7 (continued)

SOLUTION

a) The diagram shows the forces acting on particle P.

Since P is in equilibrium,

$[\uparrow]$ $T_1 - 50 = 0$

$\Rightarrow$ $T_1 = 50$

The tension in the string linking P and Q is 50 N.

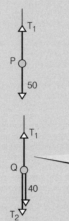

b) The second diagram shows the forces acting on Q.

Since Q is in equilibrium

$[\uparrow]$ $T_1 - T_2 - 40 = 0$

$\Rightarrow$ $T_2 = T_1 - 40$

$\Rightarrow$ $T_2 = 50 - 40 = 10$

The tension in the string linking Q and R is 10 N.

> **Modelling Assumption**
> Since the peg, A, is smooth the tension in the portion of the string linking P to A will be the same as the tension in the portion of the string linking A to Q.

> **Modelling Assumption**
> Since the peg, B, is smooth the tension in the portion of the string linking Q to B will be the same as the tension in the portion of the string linking B to R.

c) Finally, consider the forces acting on R.

$[\leftarrow]$ $T_2 - F = 0$

$\Rightarrow$ $F = T_2 = 10$

The frictional force acting on R is 10 N.

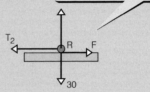

EXERCISE 2

1 A train, consisting of an engine and two trucks, travels along a straight railway track at a constant speed of 70 km/hr. The engine produces a driving force of 18 500 N and the total resistance to the motion of each truck is 4500 N.

Calculate
a) the total resistance to the motion of the engine;
b) the force in the coupling between the engine and the first truck;
c) the force in the coupling between the two trucks.

2 A small particle, A, of weight 28 N is hanging freely from one end of a string. This string passes over a small smooth fixed peg, P, which is attached to the edge of a circular table and the other end of the string is attached to a small particle, B, of weight 35 N, which lies on the table top. Another string is attached to B and this string passes over a small smooth fixed peg, Q, which is on the opposite side of the table to P. The other end of this string is attached to a small particle, C, of weight 22 N which is hanging freely.

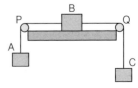

If the system is at rest, calculate
a) the tension in the string APB;
c) the normal reaction between B and the table;
b) the tension in the string BQC;
d) the frictional force acting on B.

3 A particle, P, of weight 25 N rests on a rough plane inclined at 30° to the horizontal. It is connected by a light inextensible string, passing over a small smooth fixed peg at the top of the plane, to a block, Q, of weight 35 N which is hanging freely.

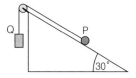

If the system is at rest, calculate
a) the tension in the string;
b) the frictional force between the particle and the inclined plane.

4 A particle P of weight 60 N is held in place on a smooth plane inclined at 15° to the horizontal by a string AP which is attached to a fixed point, A, at the top of the plane in such a way that the string is parallel to a line of greatest slope on the plane. A second particle, Q, of weight 80 N is held in place on the plane below P by a string PQ.

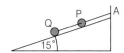

a) Draw a diagram to show the forces acting on Q. Hence find
i) the normal contact force of the plane on Q;
ii) the tension in the string PQ.
b) Draw a diagram to show the forces acting on P. Hence find
i) the normal contact force of the plane on P;
ii) the tension in the string AP.

5 Three chests, A, B and C, have weights of 200 N, 300 N and 500 N, respectively and are being pushed at a constant speed up a rough slope inclined at 12° to the horizontal by a force of 350 N acting parallel to the slope.

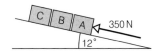

The frictional force acting on A has magnitude 30 N and the frictional force acting on B has magnitude 40 N.
a) Draw a diagram to show the external forces acting on the whole system of three crates and hence determine the magnitude of the frictional force acting on C.
b) Draw a diagram to show all the forces acting on C. Hence determine the size of
i) the normal contact force between the slope and C;
ii) the normal contact force between B and C.
c) Find the size of the normal contact force between A and B.

6 Particles K and L, each of weight 40 N, are placed on a rough plane inclined at 39° to the horizontal. K is above L. They are connected by a taut string which is parallel to a line of greatest slope of the plane. K is connected by a light inextensible string, passing over a small smooth fixed peg at the top of the plane, to a block, M, of weight 80 N which is hanging freely.

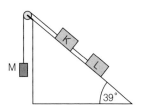

Given that the block M is descending at a constant speed and the frictional forces acting on K and L have equal magnitude, find, in any order,
i) the tension in the string KM;
ii) the frictional force acting on K;
iii) the tension in the string LK.

Modelling Frictional Forces

When two objects are in contact we know that there will be a normal reaction force acting perpendicular to the line of contact between the two objects and that there will usually be a frictional force acting along the line of contact between the two objects. This frictional force opposes any relative motion or tendency towards relative motion between the two objects. If the contact between the two objects is so well oiled that they can move relative to each other without any rubbing at all, then the friction forces can be ignored: in such cases the word "smooth" is used to describe the contact between the two objects.

The properties of friction can be illustrated by means of a simple experiment in which a sledge is initially at rest on a rough horizontal surface. The sledge is connected by a string passing over a small, smooth fixed peg to a weight which hangs freely from the string.

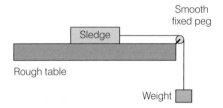

If the weight attached to the string is slowly increased from zero then the sledge will remain at rest on the table for some time but will then start to accelerate.

Thus, for a while, the frictional force between the table and the sledge is enough to prevent any motion of the sledge but, once the weight attached to the string exceeds a critical value the frictional force cannot prevent the motion of the sledge.

If

the weight of the sledge	$= S$ N
the weight attached to the string	$= P$ N
the normal reaction between the sledge and the table	$= R$ N
the frictional force between the sledge and the table	$= F$ N

then the diagrams below show the force diagrams for the sledge and for the hanging weight.

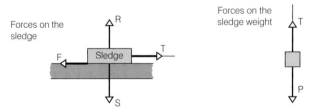

Whilst the system is at rest the forces acting on the hanging weight and the forces acting on the sledge must be in equilibrium:

$[\uparrow$, hanging weight] $\quad T - P = 0$

$\Rightarrow \quad T = P$

$[\rightarrow$, sledge] $\quad \Rightarrow \quad T - F = 0$

$\Rightarrow \quad F = T$

Putting the two results together gives

$F = P$

When the hanging weight takes the critical value, P_c, then the system will still be just in equilibrium so the result

$F = P$

is still valid. This implies that P_c, the critical value of the hanging weight, is the same as F_{max}, the maximum possible value of the frictional force between the sledge and the table.

If the hanging weight exceeds the critical value then equilibrium is broken and the sledge accelerates from rest. In this case the frictional force is usually assumed still to be at its maximum value although, in practice, it will probably be slightly less than this value.

The relationship between the frictional force and the hanging weight can usefully be shown on a graph:

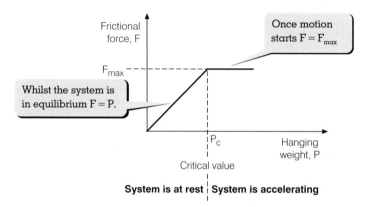

So far, the experiment leads to the conclusions that

- Friction opposes motion or the tendency towards motion but the frictional force is never more than just enough to oppose the motion.
- As the tendency towards motion increases so does the frictional force **until** it reaches a maximum value, F_{max}, which cannot be exceeded. If the tendency towards motion continues to increase then motion will start and the frictional force can be assumed to remain constant at its maximum value.

The experiment can be adapted to give a rule for F_{max}.

A load of weight L is attached to the top of the sledge and the previous experiment is repeated.

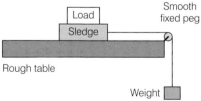

A larger hanging weight is needed to break the equilibrium of the system than was previously the case.

In particular, if the load has weight S then the critical hanging weight will be $2P_c$ and if the load has weight 2S then the critical hanging weight will be $3P_c$.

Recalling that the critical hanging weight is the same as the maximum value of the frictional force, we conclude that adding a load to the sledge increases the value of F_{max}. In particular, if the load has weight S then the maximum frictional force will be $2P_c$ and if the load has weight 2S then the maximum frictional force will be $3P_c$.

Consideration of the force diagrams will enable this to be expressed as a formula.
Let R' denote the normal reaction between the sledge and the load. The three diagrams below show the forces acting on the load, on the sledge and on the hanging weight.

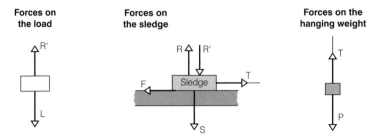

| Forces on the load | Forces on the sledge | Forces on the hanging weight |

Whilst the system is at rest,

$[\uparrow, \text{load}] \quad R' - L = 0$

$\Rightarrow \quad R' = L$ [1]

$[\uparrow, \text{sledge}] \quad R - R' - S = 0$

$\Rightarrow \quad R = R' + S$

Using [1] to substitute for R'

$\Rightarrow \quad R = L + S$

Adding a load to the sledge is changing the normal reaction between the sledge and the rough table. The results linking the load, the reaction between the sledge and the table and the maximum possible frictional force are shown in the table below.

Load, L	Normal reaction, R, between sledge and table	F_{max}
0	S	P_c
S	2S	$2P_c$
2S	3S	$3P_c$

These results clearly suggest that the maximum value of the frictional force is proportional to the reaction between the table and the sledge. We can write

$$F_{max} \propto R \quad \text{or} \quad F_{max} = \mu R$$

where μ is a constant of proportionality, called the **coefficient of friction**. The value of this constant depends on the materials that the two objects are made from, the lubrication between the two surfaces and the exact profile of the two surfaces. If the lubrication is very good, the value of μ can be very close to zero but if one of the surfaces is very sticky or very rough then the value of μ can be large.

- The maximum value of the frictional force between a particle and a surface is proportional to the normal reaction, R, and the constant of proportionality, μ, is called the coefficient of friction, μ, for the particle and the surface. Thus $F_{max} = \mu R$.

Modelling Assumption

In practice, at the microscopic level, there is no such thing as a flat even surface or a surface with a uniform profile and this means that the value of μ will vary from point to point on a surface: the adoption of a constant value of μ for all points on a surface is another example of an assumption that is usually made in the process of producing a mathematical model of a mechanical situation.

Equilibrium Examples Involving Friction

A new condition must now be included to ensure that a particle is in equilibrium: not only must the resultant force acting on the particle be zero but also the sizes of any frictional forces must be less than or equal to their maximum possible values as given by $F_{max} = \mu R$.

EXAMPLE 8

A particle of weight 20 N is placed on a rough slope inclined at 30° to the horizontal. The coefficient of friction between the slope and the particle is 0.6.

a) Show that the particle will rest in equilibrium on this slope.

A force of magnitude P N is now applied to the particle in a direction parallel to the upward direction of the slope.

b) Determine whether the particle is in equilibrium and find the size of the frictional force if
  i) $P = 18$ **ii)** $P = 24$

a)

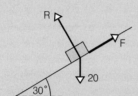

For equilibrium

[perpendicular, out] $R - 20 \cos 30° = 0$
$$\Rightarrow \quad R = 20 \cos 30° = 17.32 \dots$$

[parallel, up] $F - 20 \sin 30° = 0$
$$\Rightarrow \quad F = 20 \sin 30° = 10$$

To show that the particle is in equilibrium we must check that this frictional force does not exceed the maximum possible value:

In this case, $F_{max} = \mu R = 0.6 \times 20 \cos 30° = 10.39 \dots$ N.

Since F = 10 N and F_{max} = 10.39 ... N, the condition $F \leqslant F_{max}$ is satisfied and the particle can rest in equilibrium on the slope.

b)

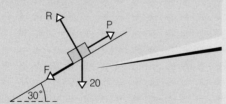

We know the component of weight down the plane has size

$$20 \sin 30° = 10 \text{ N}$$

The force P has size 18 or 24 N so the force P is enough to overcome the component of weight down the plane. The forces are trying to move the block **up** the plane. This means the friction is acting **down** the plane.

For equilibrium:

[perpendicular, out] $R - 20 \cos 30° = 0$
$$\Rightarrow \quad R = 20 \cos 30° = 17.32 \dots$$

[parallel, up] $P - F - 20 \sin 30° = 0$
$$\Rightarrow \quad F = P - 20 \sin 30°$$
$$\Rightarrow \quad F = P - 10$$

EXAMPLE 8 (continued)

However, there is a maximum possible value for the frictional force given by
$F_{max} = \mu R$.
In this case $F_{max} = \mu R = 0.6 \times 20 \cos 30° = 10.39 \ldots$ N.

i) If $P = 18$ then $F = 24 - 10 = 8$.
Since $8 < 10.39 \ldots$, the condition $F \leqslant F_{max}$ is satisfied and the particle can rest in equilibrium on the slope. The frictional force is 8 N.

ii) If $P = 24$ then $F = 24 - 10 = 14$ so a frictional force of 14 N is required for equilibrium.
However, the frictional force cannot exceed $10.39 \ldots$ N so equilibrium is not possible and the particle will start to move up the plane. In this case, the frictional force will take it largest possible value, which is 10.4 N (3 s.f.).

A particle is said to be in **limiting equilibrium** if the particle is in equilibrium and the frictional force acting on the particle is equal to the maximum possible value for the friction. A particle will be in limiting equilibrium if it is either on the point of moving or if it is moving with constant speed in a straight line.

EXAMPLE 9

A particle of weight 25 N is placed on a rough horizontal table. When the particle is pushed by a force of 20 N which is inclined at 15° below the horizontal, the particle is in limiting equilibrium. Calculate the value of μ, the coefficient of friction between the particle and the table.

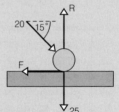

A diagram should be drawn to show the forces acting on the particle.

For equilibrium

$[\uparrow]$ $R - 25 - 20 \sin 15° = 0$

$\Rightarrow$ $R = 25 + 20 \sin 15° = 30.17 \ldots$

$[\rightarrow]$ $20 \cos 15° - F = 0$

$\Rightarrow$ $F = 20 \cos 15° = 19.31 \ldots$

Since the equilibrium is limiting,

$F = F_{max} = \mu N$

$\Rightarrow$ $\mu = \dfrac{F}{N} = \dfrac{19.31 \ldots}{30.17 \ldots} = 0.6401 \ldots$

$\Rightarrow$ $\mu = 0.640$ (3 s.f.)

EXAMPLE 10

A small block of weight 10 N rests on a rough plane inclined at 30° to the horizontal. A horizontal force of magnitude P N holds the block in equilibrium. If the coefficient of friction between the block and the plane is 0.5, show that

$$0.60 \leqslant P \leqslant 15.15$$

SOLUTION

Start by assuming that the force P is small so that the block has a tendency to move down the slope. If this is the case the frictional force will act up the slope.

For equilibrium,

[perpendicular, out] $R - P \sin 30° - 10 \cos 30° = 0$

$\Rightarrow \quad R = P \sin 30° + 10 \cos 30°$

$\Rightarrow \quad R = \dfrac{1}{2}P + 5\sqrt{3}$

> Since the angle of the slope is 30° we can use the exact values
>
> $\sin 30° = \dfrac{1}{2} \qquad \cos 30° = \dfrac{\sqrt{3}}{2}$
>
> to keep the working precise.

[parallel, up] $F + P \cos 30° - 10 \sin 30° = 0$

$\Rightarrow \quad F = 10 \sin 30° - P \cos 30°$

$\Rightarrow \quad F = 5 - \dfrac{\sqrt{3}}{2}P$

For equilibrium to be possible,

$$F \leqslant F_{max} = \mu R$$

$\Rightarrow \quad 5 - \dfrac{\sqrt{3}}{2}P \leqslant 0.5\left(\dfrac{1}{2}P + 5\sqrt{3}\right)$

$\Rightarrow \quad 5 - \dfrac{\sqrt{3}}{2}P \leqslant \dfrac{1}{4}P + \dfrac{5\sqrt{3}}{2}$

[× 4] $\Rightarrow \quad 20 - 2\sqrt{3}P \leqslant P + 10\sqrt{3}$

$\Rightarrow \quad 20 - 10\sqrt{3} \leqslant P + 2\sqrt{3}P$

$\Rightarrow \quad 20 - 10\sqrt{3} \leqslant (1 + 2\sqrt{3})P$

$\Rightarrow \quad \dfrac{20 - 10\sqrt{3}}{(1 + 2\sqrt{3})} \leqslant P$

$\Rightarrow \quad 0.60023 \ldots \leqslant P$

Now assume that the force P is large enough that the block has a tendency to move up the slope. The frictional force will now act down the slope.

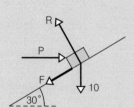

For equilibrium,

[perpendicular, out] $R - P \sin 30° - 10 \cos 30° = 0$

$\Rightarrow \quad R = P \sin 30° + 10 \cos 30°$

$\Rightarrow \quad R = \dfrac{1}{2}P + 5\sqrt{3}$

EXAMPLE 10 (continued)

[parallel, up] $P \cos 30° - F - 10 \sin 30° = 0$

$F = P \cos 30° - 10 \sin 30°$

$\Rightarrow \quad F = \frac{\sqrt{3}}{2}P - 5$

For equilibrium to be possible,

$F \leqslant F_{max} = \mu R$

$\Rightarrow \quad \frac{\sqrt{3}}{2}P - 5 \leqslant 0.5\left(\frac{1}{2}P + 5\sqrt{3}\right)$

$\Rightarrow \quad \frac{\sqrt{3}}{2}P - 5 \leqslant \frac{1}{4}P + \frac{5\sqrt{3}}{2}$

[× 4] $\Rightarrow \quad 2\sqrt{3}P - 20 \leqslant P + 10\sqrt{3}$

$\Rightarrow \quad 2\sqrt{3}P - P \leqslant 20 + 10\sqrt{3}$

$\Rightarrow \quad (2\sqrt{3} - 1)P \leqslant 20 + 10\sqrt{3}$

$\Rightarrow \quad P \leqslant \frac{20 + 10\sqrt{3}}{(2\sqrt{3} - 1)}$

$\Rightarrow \quad P \leqslant 15.145 \ldots$

Combining the two results gives $0.60 \leqslant P \leqslant 15.15$, correct to 2 decimal places.

EXERCISE 3

1 A particle of weight 10 N rests in rough contact with a plane inclined at 30° to the horizontal. Calculate the value of the coefficient of friction if the particle is just about to slip down the plane.

2 A child has a toy of weight 100 N which is attached to a string. The toy is initially at rest on rough horizontal ground. The child picks up the string and pulls it so that the string makes an angle of 16° with vertical. If the coefficient of friction between the toy and the ground is 0.8 and the toy is in limiting equilibrium, find the tension in the string.

3 A small block of weight W is placed on a plane inclined at an angle $\theta°$ to the horizontal. The coefficient of friction between the block and the plane is μ.
   a) If $\theta = 20°$ and the block is in limiting equilibrium, find the value of μ.
   b) If $\theta = 30°$ and the block is in equilbrium, prove that $\mu \geqslant \frac{1}{\sqrt{3}}$.
   c) If $\mu = \frac{1}{3}$ and $\theta = 30°$, a horizontal force of 6 N is required to prevent the block from slipping down the plane. Find the value of W.

4 A small block of weight 8 N is standing on rough horizontal ground. The coefficient of friction between the block and the ground is 0.5.
   A horizontal force P is applied to the block.
   Find the value of the frictional force and state whether the block will move in each of the following cases:
   a) P = 1 N   **b)** P = 4 N   **c)** P = 5 N

5 The diagram shows two particles A and B of weights 20 N and 30 N, respectively. A is on a rough inclined plane which makes an angle of 27° with the horizontal and is attached to one end of a light inextensible string. This string passes over a small smooth fixed peg at the top of the plane and the other end is attached to B which hangs freely. The coefficient of friction between A and the plane is μ.

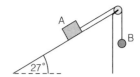

If A is sliding up the plane at constant speed, find
i)   the tension in the string;
ii)   the normal reaction between the plane and A;
iii) the frictional force between A and the plane;
iv) the value of μ.

6 A small block of weight 56 N rests on a rough horizontal table. The coefficient of friction between the table and the block is 1.5. The block is pulled by a string which is inclined at 21° above the horizontal.

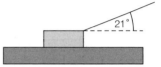

a) If the tension in the string is 50 N, prove that the block is in equilibrium.
b) If the tension in the string is 60 N, show that the block is not in equilibrium and find the size of the normal reaction and the frictional force between the block and the table.
c) Find the tension in the string if the block is in limiting equilibrium.
d) What would happen to the block if the tension in the string was 160 N?

7 A block of weight 20 N rests on a rough plane inclined at 30° to the horizontal. The coefficient of friction between the plane and the block is 0.25. Calculate the horizontal force required
a) to just prevent the block from slipping down the plane;
b) to make it just begin to slide up the plane.

8 A sledge whose weight is 4000 N is pulled at constant speed along level ground by a rope inclined at 30° to the horizontal. The coefficient of friction between the sledge and the ground is 0.25.

If R is the normal reaction between the sledge and the ground and T is the tension in the rope prove that

$$R + \frac{1}{2}T = 4000$$

and write down another equation that must be satisfied by R and T.
Find the value of T.

Suppose now the rope makes an angle θ with the horizontal. Find an expression for T in terms of θ.

Use your graphical calculator to produce a sketch showing how T varies with θ and find the minimum possible value of T and the corresponding value of θ.

Having studied this chapter you should know that

● friction opposes motion or the tendency towards motion but is never more than just enough to oppose the motion
  as the tendency towards motion increases so does the frictional force **until** it reaches a maximum value, F_{max}, which cannot be exceeded. If the tendency towards motion continues to increase then motion will start and the frictional force can be assumed to remain constant at its maximum value
  the maximum value of the frictional force between a particle and a surface is proportional to the normal reaction, R, between the particle and the surface introducing a constant of proportionality, μ, gives $F_{max} = \mu R$. μ is called the coefficient of friction for the particle and the surface

● a particle is in equilibrium if
       1) the resultant force acting on the particle is zero
  and
       2) any frictional forces satisfy $F \leqslant F_{max}$

● the equilibrium of a system of particles can be discussed by either considering the external forces acting on the whole system or by considering the forces acting on components of the system

REVISION EXERCISE

1   Find the values of P and Q if the particle is in equilibrium when the forces shown in the diagram are acting upon it:

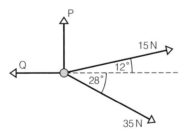

2   A Christmas decoration of weight 500 N is suspended above a road by means of two cables attached to buildings on either side of the road. One cable is horizontal whilst the other cable makes an angle of 30° with the horizontal.

Calculate the tension in each cable.

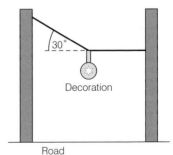

3 Two particles A and B, each of weight 20 N, are joined by a light inextensible string. Particle A rests on a rough inclined plane which is fixed at 15° to the horizontal. The string is parallel to a line of greatest slope of the plane and passes over a small smooth pulley at the top of the board. Particle B hangs vertically below the pulley. The system is in equilibrium.

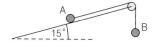

a) Calculate the tension in the string.
b) Find the frictional force acting on A.
c) If the equilibrium is limiting, with A on the point of sliding up the plane, determine the coefficient of friction between A and the plane.

4 i) Three forces of magnitude P N, 4 N and 3 N act on a particle in the directions shown in the diagram. The particle is in equilibrium. Find P and θ.

ii) The force of magnitude 4 N is now removed. The magnitudes and directions of the other two forces remain unchanged. Write down the magnitude and direction of the resultant force on the particle.

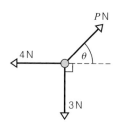

(OCR Jan 2001 M1)

5 a) A crate of weight 3000 N rests in equilibrium on a slope inclined at 20° to the horizontal. Find the frictional force acting on the crate.

b) Given that the equilibrium is limiting, calculate the coefficient of friction between the crate and the slope.

c) The crate is now pushed horizontally with a force of magnitude 2400 N as shown in the diagram. Show that the crate remains in equilibrium.

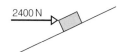

6 A particle P is attached to two points A and B by two light strings. P hangs in equilibrium with PA and PB making angles of 55° and 25°, respectively with the upward vertical, as shown in the diagram. Given that the tension in PA is 10 N, find

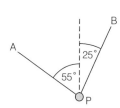

i) the tension in PB;
ii) the weight of P.

7 A lift of weight 20 000 N is moving upwards at constant speed and is carrying a load of weight 2500 N.

a) Calculate the magnitude of the normal contact force between the load and the lift floor.

b) Calculate the magnitude of the tension in the lift cable.

8 Two particles A and B of weights 5 N and w N, respectively, are joined by a light inextensible string. Particle A rests on a board which is fixed at 10° to the horizontal. The coefficient of friction between A and the board is 0.15. The string is parallel to a line of greatest slope of the board and passes over a small smooth pulley at the top of the board. Particle B hangs vertically below the pulley. The system is in equilibrium.

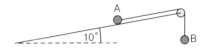

i) **a)** Given that A is on the point of sliding up the plane, show that $w = 1.61$, correct to two decimal places.

b) Given instead that A is on the point of sliding down the plane, show that $w = 0.13$, correct to two decimal places.

The board is now fixed horizontally and A is held on the board. The string passes over the pulley and B hangs vertically below the pulley. A is now released.

ii) Show that the system starts to move if w has the value in part (i)(a), but remains at rest if w has the value in part (i)(b).

iii) Find the magnitude of the frictional force acting on A for each of the values of w in part (i).

(OCR Jun 2001 M1, adapted)

9 A block of weight 20 N is placed on a horizontal table. A force of magnitude 12 N at an angle of 36° below the horizontal is applied to the block. The block remains at rest.

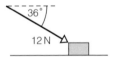

a) Find the frictional force acting on the block.

b) Find the normal component of the contact force between the block and the table.

c) Given that the block is on the point of sliding, find the coefficient of friction between the block and the plane.

10 A particle of weight 60 N rests in equilibrium on a plane inclined at 38° to the horizontal.

a) Calculate the frictional force acting on the particle.

b) Given that the equilibrium is limiting, calculate the coefficient of friction between the particle and the plane.

A force P acting up the slope is now applied to the particle, as shown in the diagram.
Determine whether or not the particle is moving and the magnitude of the frictional force in each of the cases

i) P = 60 N **ii)** P = 100 N.

11 A heavy ring of weight 8 N is threaded onto a fixed, rough horizontal rod. The coefficient of friction between the ring and the rod is $\frac{2}{5}$. A light string is attached to the ring and pulled downwards with a force of T N acting at 40° to the horizontal. Find the value of T if the equilibrium is limiting.

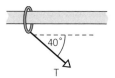

12 Particles A and B, of weight $5w$ and $3w$, respectively, resting on a smooth plane at 30° are linked by a light inextensible string parallel to the line of greatest slope of the plane. A second string joins B to a particle C of weight kw, where k is a constant. The second string is parallel to a line of greatest slope of the plane and passes over a small smooth pulley at the top of the board. Particle C hangs vertically below the pulley. The system is in equilibrium. Determine the value of k and find, in terms of w, the tension in each of the strings.

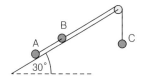

4 Kinematics 1:
   The Motion of Particles

The purpose of this chapter is to enable you to

- use distance, displacement, speed, velocity and acceleration to describe the motion of a body

- use distance–time, speed–time, displacement–time and velocity–time graphs to describe the motion of a body in a straight line

- obtain and use the constant acceleration equations for a body moving in a straight line with constant acceleration

- use the constant acceleration equations to describe vertical motion of bodies close to the Earth's surface when the only force acting on the body is its weight

Describing Motion

Distance and Displacement

The diagram shows a rectangular training area, ABCD, measuring 60 m by 40 m.

As part of his daily exercises, Nicholas jogs for ten minutes around the edge of this area at a constant speed of 5 m/s. He starts at A and runs in an anticlockwise sense.

After 15 s, Nicholas has run 75 m so he is at a point 15 m beyond B, running towards C.

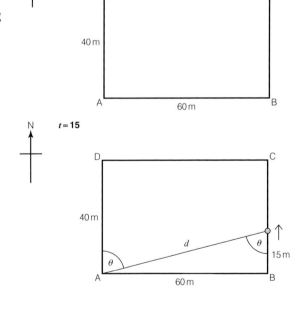

Using Pythagoras's theorem and elementary trigonometry, the distance and bearing of Nicholas from his starting point can be calculated:

$$d^2 = 60^2 + 15^2 = 3825$$
$$\Rightarrow \quad d = \sqrt{3825} = 61.84 \ldots$$

$$\tan \theta = \frac{60}{15} = 4$$

$$\Rightarrow \quad \theta = 75.96 \ldots$$

If t denotes the time, in seconds, that Nicholas has been running, then when $t = 15$

- Nicholas has run a **distance** of 75 m;
- the **displacement** of Nicholas from the starting point is 61.8 m on a bearing of 76.0°.

Similarly, when $t = 70$, Nicholas has run a distance of 350 m. This means that he has completed one full lap of the training area, has passed through both B and C for a second time, and is 50 m away from C.

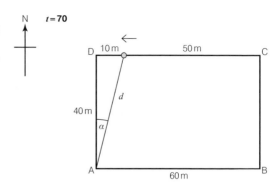

$$d^2 = 40^2 + 10^2 = 1700$$
$$\Rightarrow \quad d = \sqrt{1700} = 41.23 \ldots$$
$$\Rightarrow \quad \tan \alpha = \frac{10}{40} = 0.25$$
$$\Rightarrow \quad \alpha = 14.03 \ldots$$

So, when $t = 70$

- Nicholas has run a **distance** of 350 m;
- the **displacement** of Nicholas from the starting point is 41.2 m on a bearing of 14.0°.

Distance is a scalar quantity: it has size but no direction.
Displacement is a vector quantity which has both size and direction. A displacement can be described either as a magnitude and a direction or as a column vector.

When the motion is along a straight line, displacement can be expressed simply by using positive and negative numbers.

Consider, for example, Tara's daily exercise regime which consists of jogging along the 150 m long stretch of pavement immediately outside her house. If the pavement extends from 50 m left of the front door of Tara's house to 100 m to the right of the front door then we could represent the pavement by the portion of the number line from −50 to +100.

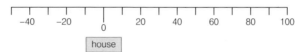

Her exercise programme consists of turning right when leaving the house and then jogging up and down this stretch of pavement at a constant speed of 3 m/s. Let t denote the time, in seconds, that Tara has been jogging.

When $t = 15$, Tara has jogged 45 m.

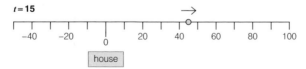

When $t = 15$, Tara has travelled a distance of 45 m and we can say that her displacement from her starting point is +45 m to show that she is 45 m away from the starting point **in the positive direction**.

When $t = 70$, Tara has jogged 210 m which means that she will have run from the house to the end of the "positive part" of the pavement, then returned to the house and continued 10 m to the left of the house.

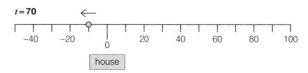

When $t = 70$, Tara has travelled a distance of 210 m and we can say that her displacement from her starting point is −10 m to show that she is 10 m away from the starting point **in the negative direction**.

Speed and Velocity

The speed of an object gives information about how quickly the object covers distances or, in other words, the rate at which the object covers distances. Unfortunately, in common usage, the word "speed" is frequently used for two different quantities. Consider the sentence

"*Yesterday Peter drove from London to Manchester at a speed of 55 miles per hour but while he was on the motorway a traffic camera recorded his speed as 82 miles per hour.*"

The figure of 55 miles per hour is really an **average speed** for the whole journey and would have been calculated using

$$\text{average speed} = \frac{\text{distance travelled}}{\text{time taken}}$$

The speed of 82 miles per hour gives the **instantaneous speed** of the car as it passes the traffic camera and this would be the figure showing on the car's speedometer at that instant. The instantaneous speed of Peter's car would obviously have varied considerably during the course of his journey from London to Manchester.

In Mechanics, the speed of a body always means the instantaneous speed of the body. If we wish to say that an object is moving at a steady speed then we shall emphasise the fact by saying that the body has **constant** speed.

The velocity of an object is the rate of change of its displacement. Since displacement is a vector quantity, velocity is also a vector quantity so has both magnitude and direction. It should be emphasised that the velocity of an object means the **instantaneous** velocity of the object.

Consider again Nicholas's exercise regime.

When $t = 15$, Nicholas is running along the edge BC of the training area. His velocity is 5 m/s due North.

When $t = 70$, Nicholas is running along the edge CD of the training area. His velocity is 5 m/s due West.

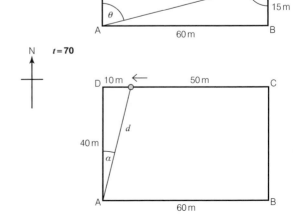

Similarly, when $t = 15$, Tara is moving at 3 m/s in the positive direction so her velocity can be written as +3 m/s.

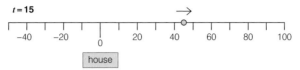

When $t = 70$, Tara is moving at 3 m/s in the negative direction so her velocity can be written as −3 m/s.

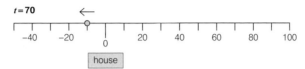

Sometimes the **average velocity** of an object over an interval of time may be needed and it can be found by calculating

$$\frac{\text{change in displacement during the interval}}{\text{length of time}}$$

For example, during the 55 second period from $t = 15$ to $t = 70$, Tara has moved from +45 to −10 so her displacement during this interval is −55 m.

Tara's average velocity from $t = 15$ to $t = 70$

$$= \frac{\text{change in displacement during the interval}}{\text{length of time}} = \frac{-55}{55} = -1 \text{ m/s}$$

Motion Graphs

Motion on a straight line can usefully be illustrated by graphs showing how the distance, displacement, speed and velocity vary with time.

Speed and Velocity Graphs

The **speed–time graph** for Tara's exercise programme shows that she runs at a constant speed of 3 m/s.

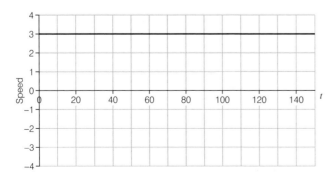

Modelling Assumption
When Tara gets to the end of the pavement she must stop and turn round to run back in the opposite direction. Drawing the speed–time graph as a straight line assumes that the turning round operation is instantaneous!

Notice that the area under the speed–time graph between $t = 0$ and $t = 70$ is $70 \times 3 = 210$, which is precisely the distance Tara has travelled in the first 70 s.

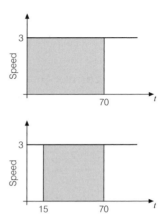

Similarly, the area under the speed–time graph between $t = 15$ and $t = 70$ is $55 \times 3 = 165$, which is also the distance she runs during these 55 s.

In general

Area under a speed–time graph represents distance travelled.

The **velocity–time graph** (also known as a (t, v) graph) for Tara's exercise programme shows that her velocity changes from +3 to −3 whenever she reaches the right-hand end of the pavement and from −3 to +3 whenever she reaches the left-hand end of the pavement.

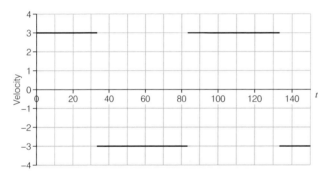

> Again, in drawing this graph we are assuming that Tara changes her direction of running instantaneously.

If the area under the velocity–time graph between $t = 0$ and $t = 15$ is calculated we obtain

$$15 \times 3 = 45$$

which is Tara's displacement from her starting position when $t = 15$.

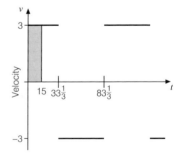

The area under the velocity–time graph between $t = 0$ and $t = 70$ must be calculated in two parts:

area **above** the t-axis $= 33\frac{1}{3} \times 3 = 100$

area **below** the t-axis $= 36\frac{2}{3} \times 3$.

> This tells us that in the first $33\frac{1}{3}$ s Tara moved 100 m in the **positive** direction.

> This tells us that in the next $36\frac{2}{3}$ s Tara moved 110 m in the **negative** direction.

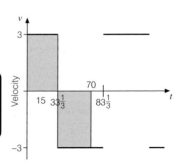

Tara has moved 100 m in the positive direction and 110 m in the negative direction, which gives a total displacement of –10 m.

In general

Displacement = (Area of velocity–time graph above *t*-axis)
– (Area of velocity–time graph below *t*-axis).

Distance and Displacement Graphs

The distance–time graph shows that the distance travelled by Tara increases at a constant rate of 3 m/s.

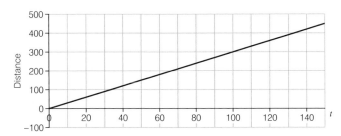

Notice that the gradient of the distance–time graph at any point on the graph is 3 m/s and this is Tara's speed.

In general

The gradient of a distance–time graph gives the instantaneous speed.

The displacement–time graph (also known as a (t, x) graph) shows Tara moving from the house to the right-hand end of the pavement and then turning round and jogging through to the left-hand end of the pavement and then turning to return again to the right-hand end of the pavement.

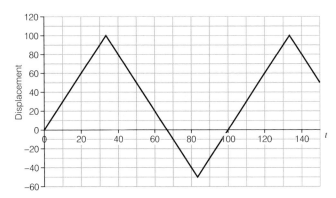

Notice that the gradient of this graph at any time gives Tara's velocity at that time: either 3 m/s or –3 m/s.

In general

The gradient of a displacement–time graph gives the instantaneous velocity.

Acceleration

A light-rail transit system links an airport, an exhibition centre and a car park.

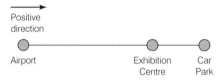

The diagram below shows the velocity–time graph for a train as it travels from the airport to the car park and then returns to the airport.

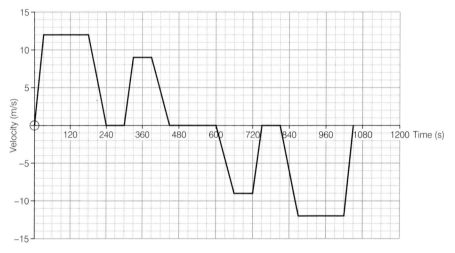

During the first 30 s, the train increases speed from 0 to 12 m/s at a constant rate.

The rate of change of velocity is called **acceleration**. Equivalently, acceleration is given by the gradient of the velocity–time graph.

We can say that during the first 30 s of the motion the train has a **constant** acceleration of $\frac{12}{30} = +0.4$ m/s^2.

For the next 150 s the train travels at a constant velocity of 12 m/s. Since the velocity is constant, the acceleration during this period is zero.

Over the next 60 s, the train's velocity decreases from 12 m/s to 0. The train is braking, presumably to come to rest at the station for the exhibition centre. During this period, the train has a constant acceleration of $\frac{-12}{60} = -0.2$ m/s^2. Alternatively, we can say that the train is braking or decelerating at a rate of 0.2 m/s^2.

The train is then at rest at the station for 60 s.

It then accelerates at a constant rate of +0.3 m/s^2 for 30 s to reach a speed of 9 m/s which is then maintained for 60 s. The train then experiences a constant acceleration of −0.15 m/s^2 for 60 s to come to rest at the station for the car park. It stays at this station for 150 s before starting its return journey.

For the first 30 s of the return journey the train has constant acceleration of −0.3 m/s². During this interval the speed of the train increases from 0 to 9 m/s but the train is moving in the negative direction. After a 60 s period of motion with a constant velocity of −9 m/s, the train brakes over 60 s to come to rest at the exhibition centre station. The acceleration during this period of braking is +0.15 m/s². (Notice that braking in the negative direction is equivalent to a positive acceleration.)

After a 60 s stop at the station the train accelerates at −0.4 m/s² for 30 s and then travels for 150 s at a constant velocity of −12 m/s. A 60 s period of +0.2 m/s² acceleration brings the train to a halt at the airport station.

Notice that if the velocity and acceleration of the train are both positive or both negative then the train's speed is increasing but if the velocity and acceleration have opposite signs then the train's speed is decreasing.

The **distances** between the stations can be calculated by finding the **areas** under relevant parts of the velocity–time graph.

Recall that the area of a trapezium with parallel sides of length a and b which are the distance h apart is given by the formula

$$\text{Area} = \frac{1}{2}(a + b)h$$

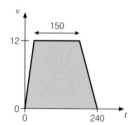

The distance between the airport and the exhibition centre is given by the area under the graph between $t = 0$ and $t = 240$:

$$\text{Distance} = \frac{1}{2} \times (240 + 150) \times 12 = 2340 \text{ m}$$

Similarly, the distance between the exhibition centre and the car park is given by the area under the graph between $t = 300$ and 450:

$$\text{Distance} = \frac{1}{2} \times (150 + 60) \times 9 = 945 \text{ m}$$

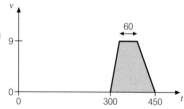

EXAMPLE 1

A motor-cyclist travels between two sets of traffic lights situated on a straight road. Starting from rest at the first set of traffic lights he accelerates to a speed of 72 km/hr and then travels for 40 s at a constant speed before braking to a halt at the second set of traffic lights. If the braking takes twice the time of the acceleration and the distance between the two sets of traffic lights is 1.16 km, find
a) the acceleration of the motor-bike away from the first set of traffic lights;
b) the average speed of the motor-bike.

EXAMPLE 1 (continued)

> Remember to work with S.I. units to ensure consistency.

$$72 \text{ km/hr} = 72 \times 1000 \text{ m/hr} = \frac{72 \times 1000}{3600} \text{ m/s}$$

$$= 20 \text{ m/s}$$

> The S.I. unit for distance is the metre, the S.I. unit for time is the second so the S.I. unit for speed and velocity is m/s or ms^{-1}.

If the time taken to accelerate to 72 km/hr is T seconds then the time taken to brake is $2T$ and the velocity–time graph for the motor-bike might be

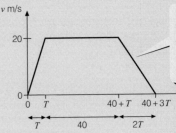

> **Modelling Assumption**
> In producing this graph we are assuming that the acceleration and braking of the motor-bike are constant. Without further information this is probably the best model that can be produced.

a) To find the acceleration, the value of T must first be determined. We know that the distance between the two sets of lights is 1160 m.

1160 = displacement of motor-cyclist as he rides between the two sets of lights

$\Rightarrow$ 1160 = area under velocity–time graph

$\Rightarrow$ $1160 = \dfrac{1}{2}((40 + 3T) + 40) \times 20$

$\Rightarrow$ $1160 = 10(80 + 3T)$

$\Rightarrow$ $116 = 80 + 3T$

$\Rightarrow$ $36 = 3T$

$\Rightarrow$ $T = 12$

The acceleration of the motor-bike is given by the gradient of velocity–time graph between $t = 0$ and $t = 12$

$$\text{acceleration } = \frac{20}{12} = 1.8333 \ldots$$

$$= 1.83 \text{ m/s}^2 \quad \text{(3 s.f.)}$$

b) For the journey between the two sets of lights

total time $= 40 + 3T = 40 + 3 \times 12 = 76$ s

The motor-bike's average speed $= \dfrac{1160}{76} = 15.26 \ldots = 15.3 \text{ m/s} \quad \text{(3 s.f.)}$

EXERCISE 1

1 Last Sunday, Michael canoed downstream from the launch site to a picnic site and then canoed upstream to the bridge.

The diagram shows a displacement–time graph for Michael's excursion, where the displacement is measured downstream from the launch site.

a) How far is the bridge from the picnic site?

b) What is Michael's speed as he canoes upstream?

c) Draw a velocity–time graph for Michael's trip in the canoe.

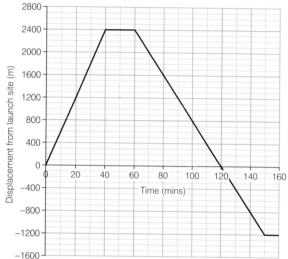

2 A department store has a basement, a ground floor and two floors above ground level. The vertical distance between one floor and the next floor is 4 m. A lift travels between the floors of the store at a speed of 0.5 m/s. When it stops at a floor, the lift waits for a minimum of 15 s before moving.

At $t = 0$, the lift is on the second floor with no passengers, and a customer on the ground floor summons the lift. The lift travels directly to the ground floor to pick the passenger up. The customer uses the lift to go to the first floor. Thirty seconds after arriving on the first floor, the lift is summoned by a customer in the basement.

Measuring displacements from the ground floor and regarding up as being the positive direction:

i) Draw, for the period of time from $t = 0$ until the lift arrives at the basement,
    a) a speed–time graph for the lift,
    b) a velocity–time graph for the lift,
    c) a distance–time graph for the lift,
    d) a displacement–time graph for the lift.

ii) Calculate, for the period of time from $t = 0$ until the lift arrives at the basement,
    a) the average speed;
    b) the average velocity.

State two assumptions that have been made about the motion of the lift in answering this question.

3 An underground train leaves a station and takes 20 seconds to reach a speed of 15 m/s. It maintains this speed for 80 seconds before braking for 10 seconds to stop at the next station.
  a) Sketch a speed–time graph for the train as it travels between the two stations.
  b) Estimate the distance between the two stations.
  c) What assumptions have you made in answering this question?

4 From a standing start at a road junction, a cyclist accelerates at a constant rate for 8 seconds to reach a speed of 27 km/hr. She then rides at a constant speed of 27 km/hr and after a total period of 40 seconds from the start of her motion she goes past a supermarket.
  a) Rewrite 27 km/hr as a speed expressed in units of m/s.
  b) Sketch the velocity–time graph of the cyclist.
  c) Calculate the acceleration of the cyclist during the first 8 seconds of her motion.
  d) Calculate the distance of the supermarket from the road junction.

5 Between seven and eight o'clock one evening, a pizza delivery boy delivered pizzas to three customers who all live on the same straight road as the pizza shop.

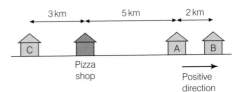

At seven o'clock the delivery boy was at the pizza shop. The diagram below shows the boy's velocity–time graph for the period from seven to eight o'clock.

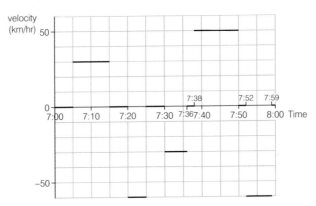

  a) Use the graph to determine the order in which the delivery boy visited the three customers.
  b) Draw a displacement–time graph to show the displacement of the delivery boy from the pizza shop for times between seven and eight o'clock that evening.

6 Peter drives anticlockwise round a 480 m long, circular go-kart track at a constant speed of 12 m/s. The start/finish line is at the most southerly point of the track. If t denotes the time, in seconds, from the time that Peter passes through the start/finish line calculate
  a) Peter's velocity when $t = 20$;
  b) Peter's average velocity between $t = 0$ and $t = 20$;
  c) Peter's velocity when $t = 30$;
  d) Peter's average velocity between $t = 20$ and $t = 30$;
  e) Peter's velocity when $t = 40$;
  f) Peter's average velocity between $t = 0$ and $t = 40$.

7 Hayley takes her small plane out for a short flight. From the airport, A, she flies 90 km due East to reach a point B, she then flies 120 km due South to reach a point C before turning to fly directly back to the airport, A. She flies the plane at a constant speed of 240 km/hr.

Calculate the velocity of Hayley's plane

a) 10 minutes after take off

b) 50 minutes after take off

c) 80 minutes after take off.

Calculate also the average velocity of Hayley's plane during the first 50 minutes of her flight.

The Constant Acceleration Equations

Suppose that a particle is moving on a straight line with constant acceleration a m/s² in such a way that initially it is at a point O on the line and moving with velocity u m/s and at time t it is at a displacement x m from O and moving with velocity v m/s.

Time 0

Time t

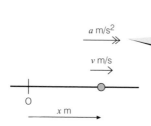

> As a convention, throughout this book a double headed arrow will be used to indicate accelerations and a single headed arrow for velocities.

Consider the velocity–time graph for the particle. We know that the acceleration, a, given by the gradient of the velocity–time graph is constant and the initial velocity of the particle is u.

If v is the velocity of the particle at time t then the (t, v) graph is a straight line of gradient a and the intercept on the velocity axis is u.

Recalling that the equation of the graph which is a straight line of gradient m and y-intercept m is

$$y = mx + c$$

we can deduce that the equation of the velocity–time graph is

$$v = at + u$$

which is usually written as

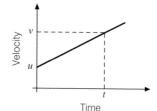

$$v = u + at$$

[1]

This is the first of five equations which, collectively, are usually known as the **constant acceleration equations**. For a particle moving with constant acceleration, these equations link

the acceleration, a
the initial velocity, u
the time, t

with

the velocity, v and the displacement, x of the particle at time t.

The displacement of the particle from O at time t is denoted by x and a formula for this can be deduced from the fact that displacement is given by the area under a velocity–time graph.

The area to be calculated is a trapezium: the parallel sides have length u and v whilst the perpendicular distance between the parallel sides is t, so

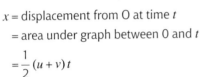

x = displacement from O at time t

= area under graph between 0 and t

$$= \frac{1}{2}(u + v)t$$

So

$$x = \frac{1}{2}(u + v)t \qquad [2]$$

Using equation [1] to substitute for v in [2] gives

$$x = \frac{1}{2}(u + u + at)t$$

$$\Rightarrow \quad x = \frac{1}{2}(2u + at)t$$

$$\Rightarrow \quad x = ut + \frac{1}{2}at^2$$

This is the third of the constant acceleration equations:

$$x = ut + \frac{1}{2}at^2 \qquad [3]$$

Equation [1] can be rewritten as

$$u = v - at$$

and substituting this into equation [2] gives

$$x = \frac{1}{2}(v - at + v)t$$

$$\Rightarrow \quad x = \frac{1}{2}(2v - at)t$$

$$\Rightarrow \quad x = vt - \frac{1}{2}at^2$$

$$x = vt - \frac{1}{2}at^2 \qquad [4]$$

Rearranging [1] to make t the subject of the formula gives

$$at = v - u$$

$$\Rightarrow \quad t = \frac{v - u}{a}$$

and substituting this into equation [2] gives

$$x = \frac{1}{2}(u + v)\frac{(v - u)}{a}$$

$$[\times 2a] \quad \Rightarrow \quad 2ax = (u + v)(v - u)$$

$$\Rightarrow \quad 2ax = v^2 - u^2$$

$$\Rightarrow \quad v^2 = u^2 + 2ax$$

which is the last of the five constant acceleration equations:

$$v^2 = u^2 + 2ax \qquad\qquad [5]$$

Notation:

It is usual to use the letters

t	to denote time
a	to denote acceleration
u	to denote the initial velocity
v	to denote the final velocity
x or s	to denote the displacement from the initial position at time t

The constant acceleration equations can therefore be written

$$v = u + at$$

$$x = ut + \frac{1}{2}at^2 \qquad \text{or} \qquad s = ut + \frac{1}{2}at^2$$

$$x = vt - \frac{1}{2}at^2 \qquad \text{or} \qquad s = vt - \frac{1}{2}at^2$$

$$x = \frac{1}{2}(u + v)t \qquad \text{or} \qquad s = \frac{1}{2}(u + v)t$$

$$v^2 = u^2 + 2ax \qquad \text{or} \qquad v^2 = u^2 + 2as$$

Although you will probably use the equations

$$v = u + at$$

and

$$x = ut + \frac{1}{2}at^2$$

most often, all five of the equations need to be learnt!

EXAMPLE 2

A cyclist accelerates steadily from rest at 0.8 m/s^2. Find

a) her speed after 5 s;

b) the distance she has travelled by the time her speed reaches 7.5 m/s;

c) the time she takes to travel 24 m.

EXAMPLE 2 (continued)

a) $a = 0.8$
$u = 0$
$t = 5$
$v = ?$
$x = ✗$

> A quick list showing the known values of a, u, t, v and x, the values to be calculated and the values that are of no interest can often help in deciding which equation to use. We will use "?" to indicate values to be calculated and "✗" to indicate values that are of no interest.

> In this case we do not know and are not interested in the value of x, so we use the equation which does not mention x:
> $$v = u + at$$

$v = u + at = 0 + 0.8 \times 5 = 4$
After 5 s, the cyclist's speed is 4 m/s.

b) $a = 0.8$
$u = 0$
$t = ✗$
$v = 7.5$
$x = ?$

> In this case we do not know and are not interested in the value of t, so we use the equation which does not mention t:
> $$v^2 = u^2 + 2ax$$

$$v^2 = u^2 + 2ax$$
$$\implies 7.5^2 = 0^2 + 2 \times 0.8 \times x$$
$$\implies 56.25 = 1.6x$$
$$\implies x = 35.15625$$
$$\implies x = 35.2 \quad (3 \text{ s.f.})$$

The distance travelled by the time her speed is 7.5 m/s is 35.2 m.

c) $a = 0.8$
$u = 0$
$t = ?$
$v = ✗$
$x = 24$

> In this case we do not know and are not interested in the value of v, so we use the equation which does not mention v:
> $$x = ut + \frac{1}{2}at^2$$

$$x = ut + \frac{1}{2}at^2$$
$$\implies 24 = \frac{1}{2} \times 0.8 \times t^2$$
$$\implies 24 = 0.4t^2$$
$$\implies t^2 = 60$$
$$\implies t = \sqrt{60} = 7.7459 \dots$$
$$\implies t = 7.75 \text{ s} \quad (3 \text{ s.f.})$$

> Ignore $t = -\sqrt{60}$ since negative times have no significance in this problem.

The time taken by the cyclist to ride 24 m is 7.75 s.

EXAMPLE 3

The brakes on a bicycle produce a retardation of 4 m/s^2. Find

a) the time to stop from a velocity of 8 m/s;
b) the distance required to stop from a speed of 12 m/s.

SOLUTION

Earlier in the chapter, we observed that if the velocity and acceleration of an object have the same sign (are in the same direction) then the speed of the object increases but if they have different signs (are in opposite directions) then the speed of the object decreases. The velocity and acceleration of a braking bicycle must therefore be in opposite directions.

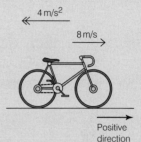

4 m/s^2

8 m/s

Positive direction

a) $a = -4$
$u = 8$
$t = ?$
$v = 0$
$x = ✗$

The acceleration is 4 m/s^2 in the **negative** direction so $a = -4$.

$v = u + at$

In this case we do not know and are not interested in the value of x, so we use the equation which does not mention x:

$v = u + at$

$\Rightarrow \quad 0 = 8 + (-4)t$

$\Rightarrow \quad 4t = 8$

$\Rightarrow \quad t = 2$

The time to stop from a velocity of 8 m/s is 2 s.

b) $a = -4$
$u = 12$
$t = ✗$
$v = 0$
$x = ?$

In this case we do not know and are not interested in the value of t, so we use the equation which does not mention t:

$v^2 = u^2 + 2ax$

$v^2 = u^2 + 2ax$

$\Rightarrow \quad 0^2 = 12^2 + 2 \times (-4) \times x$

$\Rightarrow \quad 0 = 144 - 8x$

$\Rightarrow \quad 8x = 144$

$\Rightarrow \quad x = 18$

The distance required to stop from a speed of 12 m/s is 18 m.

EXAMPLE 4

A train travelling along a straight line with constant acceleration is observed to cover consecutive distances of 1 km in times of 60 s and 40 s, respectively.
Find the initial velocity and acceleration of the train.

A diagram will help to show all the information.

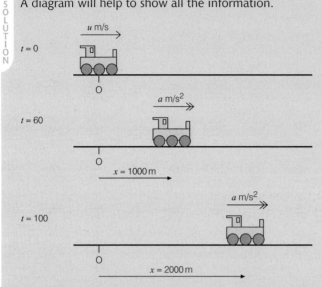

When $t = 60$ we have

$a = ?$
$u = ?$
$t = 60$
$v = \textbf{X}$
$x = 1000$

In this case we do not know and are not interested in the value of v, so we use the equation which does not mention v:

$$x = ut + \frac{1}{2}at^2$$

The equation

$$x = ut + \frac{1}{2}at^2$$

gives

$$1000 = u \times 60 + \frac{1}{2} \times a \times 60^2$$

$$\Rightarrow \quad 60u + 1800a = 1000$$

A similar equation linking u and a can be obtained using the information for $t = 100$:

$a = ?$
$u = ?$
$t = 100$
$v = \textbf{X}$
$x = 2000$

The equation

$$x = ut + \frac{1}{2}at^2$$

EXAMPLE 4 (continued)

gives

$$2000 = u \times 100 + \frac{1}{2} \times a \times 100^2$$

$$\Rightarrow \quad 100u + 5000a = 2000$$

This gives two simultaneous equations for the unknowns u and a:

$$\left.\begin{array}{l} 60u + 1800a = 1000 \\ 100u + 5000a = 2000 \end{array}\right\} \Rightarrow u = 11.666\ldots, a = 0.1666\ldots$$

The solution to the simultaneous equations can be found algebraically or by using a graphical calculator.

The initial velocity of the train is 11.7 m/s (3 s.f.) and the acceleration is 0.167 m/s^2 (3 s.f.)

EXAMPLE 5

A and B are two points 200 m apart on a straight line.
A particle, P, starts from A with initial velocity 5 m/s directed towards B and travels along a straight line with constant acceleration 2 m/s^2 directed towards B. Two seconds later a second particle, Q, starts from rest at B and travels along the same line with an acceleration of 6 m/s^2 directed towards A.
Determine when and where the two particles collide.

Let t denote the time, in seconds, that has elapsed from the instant when P started to move.
The diagram shows the positions of P and Q at time t.

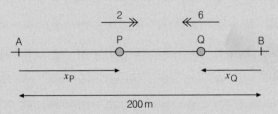

The equation $x = ut + \frac{1}{2}at^2$ gives the displacement of P from A at time t:

$$x_P = 5t + \frac{1}{2} \times 2 \times t^2$$

$$\Rightarrow \quad x_P = 5t + t^2$$

The equation $x = ut + \frac{1}{2}at^2$ will give the displacement of Q from B at time t provided we remember that, at time t, Q has only been moving for $(t - 2)$ seconds since the motion of Q commenced 2 s after the motion of P started.

$$x_Q = 0(t - 2) + \frac{1}{2} \times 6 \times (t - 2)^2$$

$$\Rightarrow \quad x_Q = 3(t - 2)^2$$

EXAMPLE 5 (continued)

The particles meet when

$$x_P + x_Q = 200$$
$$\Rightarrow \quad 5t + t^2 + 3(t-2)^2 = 200$$
$$\Rightarrow \quad 5t + t^2 + 3(t^2 - 4t + 4) = 200$$
$$\Rightarrow \quad 5t + t^2 + 3t^2 - 12t + 12 = 200$$
$$\Rightarrow \quad 4t^2 - 7t - 188 = 0$$
$$\Rightarrow \quad t = 7.78626 \ldots \text{ or } -6.03626 \ldots$$

> Solved by graphical calculator or by the formula for quadratic equations.

The value of t must be positive since the motion starts when $t = 0$

$$\Rightarrow \quad t = 7.78626 \ldots$$

and substituting this value of t into the expression for x_P gives

$$x_P = 5 \times 7.78626 \ldots + (7.78626 \ldots)^2 = 99.557 \ldots$$

So, the particles collide after P has been moving for 7.79 s (3 s.f.) and at a point which is 99.6 m (3 s.f.) from A.

EXERCISE 2

1. A particle with an initial velocity of 8 m/s moves in a straight line with a constant acceleration of 1.5 m/s^2 for 6 seconds. Find the final velocity and the distance covered.

2. A particle is moving in a straight line with a constant acceleration of −2 m/s^2. If the initial velocity is 16 m/s
   a) calculate the velocity of the particle when $t = 6$, 8 and 10;
   b) sketch a velocity–time graph for the first 10 seconds of the particle's motion;
   c) calculate the displacement of the particle from its initial position when $t = 6$, 8 and 10;
   d) sketch a displacement–time graph for the first ten seconds of the particle's motion.

3. A particle which is moving in a straight line with constant acceleration 1.6 m/s^2 is initially at a point O on the straight line and moving with velocity 4 m/s. Find
   a) the displacement of the particle from O after 3 seconds of motion;
   b) the displacement of the particle from O after 4 seconds of motion;
   c) the displacement of the particle during the fourth second of the motion.

4. P and Q are two points on a straight line with PQ = 12.8 m. A particle moves on this straight line with constant acceleration. Initially the particle is at P and 2 seconds later the particle passes through Q with a velocity of 4.8 m/s heading away from P. Find the magnitude and direction of the initial velocity and acceleration of the particle.

5. A particle moving in a straight line with a constant acceleration of −1.5 m/s^2 has an initial velocity of 18 m/s. Find
   a) when the velocity is zero;
   b) when it returns to its starting point;
   c) the distance travelled by the particle before returning to its starting point.

6 A particle moves along a straight line with constant acceleration 1.4 m/s^2. It passes though a point, A, with velocity 3 m/s and later passes through a second point, B, with velocity 10 m/s. Calculate the distance AB.

7 A particle moving in a straight line with constant acceleration has a velocity of 12 m/s at a particular instant. At a later instant the velocity is 18 m/s. The particle has moved 30 m between the two instants. Find the acceleration of the particle and the time between the two instants.

8 A particle starts from rest at a point O on a straight line and moves along the line with a constant acceleration of 3.2 m/s^2. Five seconds later a second particle passes through O with velocity 3 m/s and moves along the same line with constant acceleration 5.4 m/s^2. Find when the second particle overtakes the first and the distance of the two particles from O when this occurs.

9 Two particles are travelling along a straight line AB of length 70 m.
At the same instant one particle passes through A with velocity 6 m/s directed towards B and travels towards B with a constant acceleration of 2 m/s^2 and the other particle starts from rest at B and travels towards A with a constant acceleration of 6 m/s^2. Find how far from A the particles collide.

10 A cyclist rides along a straight road. She passes three points A, B and C, in that order, where AB = 130 m and BC = 120 m. She passes B 10 seconds after she passed A and she passes C a further 15 seconds later. Assuming that the cyclist is travelling with constant acceleration, find
a) the speed of the cyclist as she passes A and her acceleration;
b) the distance AD, where D is the point where the cyclist comes to rest.

11 Two speed cameras are situated 600 m apart on a straight road. A car passes the first camera at a speed of 70 km/hr and passes the second camera at a speed of 85 km/hr. Stating any assumptions made, estimate the acceleration of the car and the time that the car takes to travel between the two speed cameras.

12 Given that a car travelling at a speed of 80 km/hr has a braking distance of 37.5 m, find the retardation produced by the car's brakes, giving your answer in m/s^2, and deduce the braking distance for the same car travelling at a speed of 110 km/hr. State any assumptions made in answering this question.

13 A downhill skier travels down a slope in a straight line with constant acceleration. She passes close to four trees, T_1, T_2, T_3 and T_4. She passes T_2 10 seconds after passing T_1 and the distance between these trees is 90 m; she passes T_3 15 seconds after passing T_2 and the distance between these trees is 172.5 m. If the distance between T_3 and T_4 is 140 m, find the time that elapses from when she passes T_1 to when she passes T_4 and her speed as she passes T_4.

Vertical Motion Under Gravity

Galileo (1564–1642) discovered that a body dropped at any point near the Earth's surface experiences a constant downward acceleration of approximately 9.8 m/s^2 provided the air resistance on the body was negligible.

In fact, there are slight geographical variations in the magnitude of this acceleration due to the fact that the Earth is not a perfectly uniform sphere. Moreover, the acceleration due to gravity does decrease with the altitude of the body above the Earth's surface and the table illustrates this variation.

Altitude above Earth's surface (km)	Downward acceleration (m/s^2)
0	9.80
1	9.80
10	9.77
100	9.50
1000	7.32
10 000	1.49

From this table it can be seen that if the maximum altitude of the body is less than 10 km then a model assuming a constant gravitational acceleration of 9.8 m/s^2 will be reliable.

> If an object is moving vertically close to the Earth's surface
>
> and the only appreciable force acting on the body is its weight – in particular the air resistance is negligible
>
> and the altitude of the body above the Earth's surface does not exceed 10 km
>
> then the body experiences a constant downward acceleration of 9.8 m/s^2 and the motion of the body can be modelled using the constant acceleration equations.

EXAMPLE 6

A ball is thrown vertically upwards with speed 20 m/s from the top of a tower of height 25 m. Find

a) the time until the ball hits the ground;
b) the speed with which the ball hits the ground.

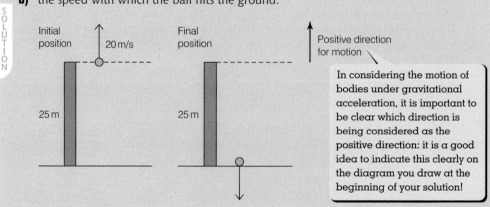

In considering the motion of bodies under gravitational acceleration, it is important to be clear which direction is being considered as the positive direction: it is a good idea to indicate this clearly on the diagram you draw at the beginning of your solution!

EXAMPLE 6 (continued)

If we take **up** as our positive direction then

a) $a = -9.8$ — The acceleration is 9.8 m/s² **down** so, since up is positive, $a = -9.8$.
$u = 20$
$t = ?$
$v = ✗$ — The final position is 25 m **below** the initial position so, since up is positive, $x = -25$.
$x = -25$

$$x = ut + \frac{1}{2}at^2$$

In this case we do not know and are not interested in the value of v, so we use the equation which does not mention v:

$$x = ut + \frac{1}{2}at^2$$

$$\implies \quad -25 = 20t + \frac{1}{2} \times (-9.8) \times t^2$$

$$\implies \quad -25 = 20t - 4.9t^2$$

$$\implies \quad 4.9t^2 - 20t - 25 = 0$$

Obtain the solutions of this quadratic by using the quadratic equation formula or a graphical calculator.

$$\implies \quad t = 5.08498 \ldots \text{ or } -1.00335$$

Since the motion is only defined for positive values of t, the negative solution can be ignored:

$$\implies \quad t = 5.08498 \ldots$$

The ball hits the ground after 5.08 s (3 s.f.).

b) Using $v = u + at$ gives

$$v = 20 + (-9.8) \times 5.08498 \ldots = -29.83 \ldots$$

i.e., the final velocity of the ball is 29.83 … **downwards**.

so the speed of the ball as it hits the ground is 29.8 m/s (3 s.f.).

Notice that we do **not** need to consider the upward and downward motion of the ball separately since the acceleration of the ball is constant **throughout** the motion and the constant acceleration equations can therefore be used for the **whole** motion.

EXAMPLE 7

A stone is thrown vertically upwards from the top of a cliff with initial speed 8 m/s and later hits the sea below with speed 17 m/s. Find the height of the cliff.

Again, a diagram should be drawn to illustrate the example.

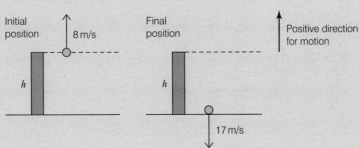

73

EXAMPLE 7 (continued)

If we take **up** as our positive direction then

$a = -9.8$

$u = 8$

$t = \mathsf{X}$

$v = -17$

$x = -h = ?$

> In this case we do not know and are not interested in the value of t, so we use the equation which does not mention t:
> $$v^2 = u^2 + 2ax$$

$$v^2 = u^2 + 2ax$$

$\Rightarrow \quad (-17)^2 = 8^2 + 2 \times (-9.8) \times (-h)$

$\Rightarrow \quad 289 = 64 + 19.6h$

$\Rightarrow \quad 19.6h = 225$

$\Rightarrow \quad h = 11.479 \ldots$

The cliff is 11.5 m high. (3 s.f.)

EXERCISE 3

1 A stone is dropped from rest from the top of a tower 80 m high. Find the speed with which it hits the ground.

2 A ball is thrown vertically upwards with a speed of 15 m/s. Find the greatest height reached by the stone.

3 A particle is projected vertically upwards from ground level with a speed of 28 m/s.
   a) Sketch a velocity–time graph for the particle.
   b) Show that the highest point reached by the particle is 40 m above ground level.
   c) Sketch a displacement–time graph for the particle.
   d) Find the length of time for which the particle is more than 10 m above ground level.

4 A stone is thrown vertically upwards at 14 m/s from the top of a cliff which is 56 m above sea level. Find the time at which the stone hits the sea and its speed at this instant.

5 At the same time as a stone is dropped from rest from a point A at the top of a 120 m high cliff a ball is thrown vertically up with speed 30 m/s from a point B which is at the bottom of the cliff and immediately below A. Find the height above B of the point where the ball and the stone collide. What assumptions have you made in answering this question?

6 A particle is projected vertically with initial speed 120 m/s from a point on ground level close to a block of flats. Vicky has a window in her flat: the bottom of the window is 35 m above ground level and the window is 3 m high. How long does the rocket take to go past Vicky's window?

Having studied this chapter you should know

● that displacement, velocity and acceleration are vector quantities that have both magnitude and direction and that directions for motion in a straight line can be expressed using + and − signs

● how to use velocity–time and displacement–time graphs to illustrate motion in a straight line

● that for motion in a straight line
   1. the gradient of a displacement–time or (t, x) graph gives velocity
   2. the gradient of a velocity–time or (t, v) graph gives acceleration
   3. the area under a velocity–time or (t, v) graph gives displacement

● the constant acceleration equations for motion in a straight line with **constant** acceleration a:

$$v = u + at$$
$$x = ut + \frac{1}{2}at^2$$
$$x = vt - \frac{1}{2}at^2$$
$$x = \frac{1}{2}(u + v)t$$
$$v^2 = u^2 + 2ax$$

and be able to use these equations in appropriate examples

● that if an object is moving vertically close to the Earth's surface and the only force acting on the object is its weight then the object has a constant downward acceleration of 9.8 m/s²

REVISION EXERCISE

1 A particle P travels in a straight line with constant acceleration 0.5 m/s².
The initial speed of P is 3 m/s. Find
a) the speed of P after 4 s;
b) the time taken for P to travel a distance of 55 m.

(OCR Jan 2001 M1)

2 A ball is at the point O on a snooker table when it is set in motion along the length of the table towards the point A at the end of the table. When the ball reaches A it rebounds along the line AO and comes to rest at B.

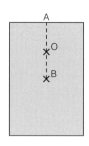

The (t, v) graph for the motion is shown in the diagram below.

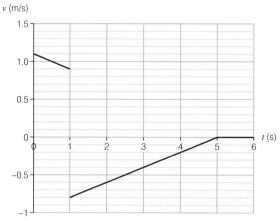

i) Find the distance OA.

ii) Find the distance OB.

iii) Show that the deceleration of the ball from O to A is the same as that from A to B.

(OCR Jun 2001 M1)

3 A car can accelerate from 0 to 100 km/hr in 8 seconds. Estimate, making your assumptions clear, the acceleration of the car in m/s^2.

4 At $t = 0$ a car passes an observer at a point O. The car is travelling with speed 20 m/s as it passes O but is braking at 0.4 m/s^2.
Three seconds later another car passes O. This car is travelling with constant speed 25 m/s.

a) Prove that the distance between the two cars at time t (where $t > 3$) is d metres where

$$d = 75 - 5t - 0.2t^2$$

b) Find when the cars meet and the distance from O when they meet.

5 The diagram shows a sketch of the (t, v) graph for the first 15 seconds of the motion of a particle P. The graph consists of two straight line segments. The particle starts at a point A and moves in a straight line.

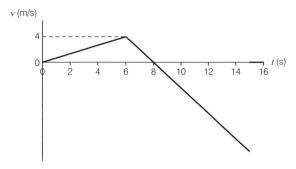

i) Calculate the displacement of P from A when $t = 8$.
ii) Find the acceleration for $t > 6$.
iii) Calculate the speed of P when $t = 15$.
iv) Show that when $t = 12$, P is at A.

(OCR Jan 2002 M1)

6 A bus leaves a bus stop and takes 30 seconds to reach a speed of 63 km/hr. It maintains this speed for 150 seconds before braking for 10 seconds to stop at the next bus stop.
a) Sketch a speed–time graph for the bus as it travels between the two stops.
b) Estimate the distance between the two bus stops.
c) What assumptions have you made in answering this question?

7 A particle P moves on a straight line with constant acceleration. At $t = 0$ P passes through a point O on the line. When $t = 20$ P passes through a point A where OA = 220 m and when $t = 50$ P passes through a point B where OB = 250 m and AB = 30 m.
a) Calculate the initial velocity and acceleration of P.
b) Determine the time when P is at rest and its distance from O at this time.

8 A stone is to be thrown upwards with speed 16 m/s from the top of a tower which is 35 metres above the surrounding ground.
a) How long will the stone take to hit the ground?
b) With what speed will the stone hit the ground?
c) What assumptions have you made in answering this question?

9 A train takes 5 minutes to cover 3 km between stations P and Q. Starting from rest at P, it accelerates at a constant rate to a speed of 40 km/hr and maintains this speed for a while and then is brought uniformly to rest at Q. If the braking takes three times as long as the acceleration, find the time taken for the train to accelerate.

10 A particle is projected vertically upwards, from the ground, with a speed of 28 m/s. Ignoring air resistance, find
a) the maximum height reached by the particle;
b) the speed of the particle when it is 30 m above the ground;
c) the time taken for the particle to fall from its highest point to a height of 30 m;
d) the length of time for which the particle is more than 30 m above the ground.

(OCR May 2002 M1)

11 Two points A and B are 64 metres apart on a straight line.

At a certain instant a particle P leaves A with initial velocity 4 m/s in the direction $\overrightarrow{AB}$ and moves with constant acceleration of 2 m/s^2 in the direction $\overrightarrow{AB}$.

At the same instant a particle Q leaves B with initial velocity 2 m/s in the direction $\overrightarrow{BA}$ and moves with constant acceleration of 3 m/s^2 in the direction $\overrightarrow{BA}$.

a) Write down expressions for

  i) the distance of P from A t seconds after P left A;

  ii) the distance of Q from B t seconds after Q left B.

b) Calculate the time when P and Q collide.

c) Find the velocities of P and Q just before the collision.

12 A woman runs from A to B, then from B to A and then from A to B again, on a straight track, taking 90 s. The woman runs at constant speed throughout. The diagram shows the (t, v) graph for the woman.

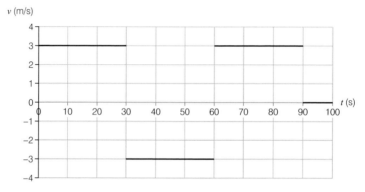

i) Find the total distance run by the woman.

ii) Find the distance of the woman from A when $t = 50$ and when $t = 80$.

At $t = 0$, a child also starts to move from A, along AB. The child walks at a constant speed for the first 50 s and then at an increasing speed for the next 40 s. The second diagram shows the (t, v) graph for the child: it consists of two straight line segments.

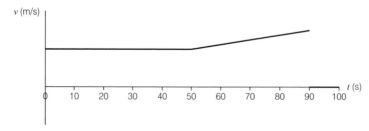

iii) At time $t = 50$, the woman and child pass each other, moving in opposite directions. Find the speed of the child during the first 50 s.

iv) At time $t = 80$, the woman overtakes the child. Find the speed of the child at this instant.

(OCR May 2002 M1)

5 Newton's Laws of Motion

The purpose of this chapter is to enable you to

- apply Newton's laws to the motion of a body in a straight line

- apply Newton's laws to the motion of two or more connected particles

Newton's three laws of motion form the basis of all the Mechanics you will study at "AS" and "A" level. Although they are only experimental laws (rather than mathematical theorems) their value comes from the fact that they give mathematical models that accurately predict the motion of most objects. The model only fails when consideration is given to the motion of very small or very large objects and when consideration is given to bodies moving at speeds approaching the speed of light.

Newton's First Law

We have already made significant use of Newton's first law of motion in chapter 3 where the equilibrium of particles was studied.

> **The first law** states that
> **"A particle will remain at rest or continue to move with constant velocity unless there is a non-zero resultant force acting on it."**

Thus
  if a particle is accelerating then there is a resultant force
and
  if a particle is not accelerating then there is no resultant force acting on the particle so the particle is in equilibrium.

EXAMPLE 1

The diagram shows the forces acting on a particle.
Is the particle accelerating?

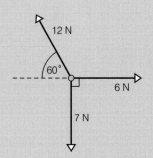

The resultant force of the system must be found: if it is non-zero then the system is accelerating.
Let X and Y denote the horizontal and vertical components of the resultant force:

$$[\rightarrow] \quad X = 6 + (-12 \cos 60°) = 0$$
$$[\uparrow] \quad Y = 12 \sin 60° + (-7) = 3.392\ldots$$

The resultant force is non-zero so the particle is accelerating.

EXAMPLE 2

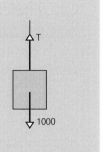

A lift of weight 1000 N is supported by a cable. The lift moves up at a constant 2 m/s. Find the tension in the cable.

The lift is moving with constant velocity so it has no acceleration.

Newton's first law implies that the resultant force must therefore be zero:

$$[\uparrow] \qquad T - 1000 = 0$$
$$\Rightarrow \quad T = 1000$$

The tension in the cable is 1000 N.

Newton's Second Law

Newton's second law is the single most important result in Mechanics since it links the motion of particles to the forces acting on a particle. More precisely, if the forces acting on a particle are known then Newton's second law allows the magnitude and direction of the acceleration of the particle to be readily deduced.

The second law states that
"When an external resultant force is applied to a body, there is an acceleration whose magnitude is proportional to the resultant force and which is in the same direction as the resultant force."

Consider what this tells us about the forces and motion of a block being pulled across a perfectly smooth table by a horizontal force P:

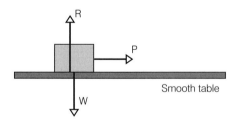

Smooth table

- Since the motion is horizontal, the acceleration is horizontal so the resultant force must be horizontal. The vertical component of the resultant force must therefore be zero:

$$[\uparrow] \qquad R - W = 0$$
$$\Rightarrow \quad R = W$$

- The resultant of the system of forces is P horizontally. The size of the acceleration is proportional to the resultant force so if the pulling force is doubled, the acceleration will double and if the pulling force is trebled, the acceleration will be trebled.

In general, if F denotes the resultant force acting on the particle of mass m and a denotes the particle's acceleration then Newton's second law states that

$$a \propto F \quad \text{and} \quad F \text{ is in the same direction as } a$$

Experiments with a constant resultant force acting on bodies of different masses suggest that the acceleration is inversely proportional to the mass:

$$a \propto \frac{1}{m}$$

Combining these two proportionality results gives

$$a \propto \frac{F}{m}$$

and, introducing a constant of proportionality, this can be rewritten as

$$ka = \frac{F}{m}$$

Rearranging to make F the subject gives

$$F = kma$$

where the value of the constant k will depend on the units being used.

Throughout this book we have been using S.I. units in calculations so mass is measured in kilograms and acceleration is measured in m/s^2. We have been using the Newton as our unit for force but so far no proper definition of a Newton has been given.

Formally, one Newton is defined to be the force required to give a particle of mass 1 kg an acceleration of 1 m/s^2.

So if we use S.I. units when using the equation $F = kma$, we know that $m = 1$ and $a = 1$ gives $F = 1$ so

$$F = kma \quad \text{gives} \quad 1 = k \times 1 \times 1 \Rightarrow k = 1$$

and we have the final mathematical formulation of Newton's second law:

$$F = ma$$

Remember that

- F is the resultant of the system of forces;
- this is a vector equation so the directions of the resultant force and the acceleration are the same;
- S.I. units must be used. The force must be measured in Newtons, the mass in kg and the acceleration in m/s^2.

The Connection Between Mass and Weight

As an immediate consequence of Newton's second law, consider a particle of mass m kg just after it has been released from rest close to the Earth's surface.

The only force acting is its weight. (Since it has only just been released, its speed will be very low so the air resistance can be assumed to be negligible.)

From our earlier work we know that the particle will have a constant downward acceleration of g ($\approx$9.8) m/s^2.

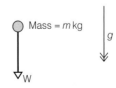

Newton's second law states that

resultant force = mass × acceleration

so

$$W = mg$$

In other words,

A body of mass m kg has a weight of mg N

Using Newton's Second Law

Since Newton's second law is a vector equation, we can apply the equation in any direction and say that

**Component of the resultant force in a given direction
= mass × component of the acceleration in the given direction.**

Confident application of this principle lies at the heart of much work in Mechanics.

EXAMPLE 3

A block of mass 20 kg moves across a smooth horizontal surface under the action of two constant horizontal forces: the first has magnitude 10 N and the second acts in the opposite direction and has magnitude 4 N. Find the acceleration of the particle and the distance covered by the particle as it accelerates from rest to a speed of 1.8 m/s.

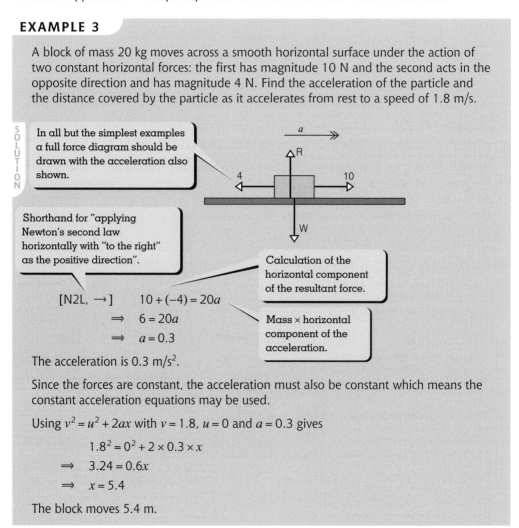

In all but the simplest examples a full force diagram should be drawn with the acceleration also shown.

Shorthand for "applying Newton's second law horizontally with "to the right" as the positive direction".

Calculation of the horizontal component of the resultant force.

$$[\text{N2L, } \rightarrow] \qquad 10 + (-4) = 20a$$
$$\Rightarrow \quad 6 = 20a$$
$$\Rightarrow \quad a = 0.3$$

Mass × horizontal component of the acceleration.

The acceleration is 0.3 m/s².

Since the forces are constant, the acceleration must also be constant which means the constant acceleration equations may be used.

Using $v^2 = u^2 + 2ax$ with $v = 1.8$, $u = 0$ and $a = 0.3$ gives

$$1.8^2 = 0^2 + 2 \times 0.3 \times x$$
$$\Rightarrow \quad 3.24 = 0.6x$$
$$\Rightarrow \quad x = 5.4$$

The block moves 5.4 m.

For problems involving inclined planes, the acceleration will usually be either up or down the plane so it is usual to apply Newton's second law in directions parallel and perpendicular to the plane.

The components of the weight in these two directions will always have to be found. The techniques of chapter 2 can continue to be used or a generalised result can be derived.

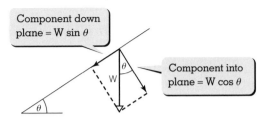

Component down plane = W sin θ

Component into plane = W cos θ

From the diagram we can see that if a particle of weight W lies on a plane inclined at θ to the horizontal then the weight W can be split into

a component down the plane of magnitude W sin θ

and

a component into the plane of magnitude W cos θ

EXAMPLE 4

A particle of mass 5 kg slides down a smooth plane inclined at 30° to the horizontal. Find

a) the acceleration of the particle;
b) the reaction between the particle and the plane.

The particle has a weight of $5g$ N. A diagram showing all the forces and the acceleration of the particle must be drawn.

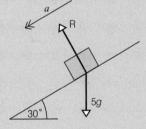

Shorthand for "applying Newton's second law parallel to the plane with "down" as the positive direction".

Calculation of the component down the plane of the resultant force.

a) [N2L, parallel, down] $5g \sin 30° = 5a$

Mass × component down the plane of the acceleration.

$\Rightarrow \quad 24.5 = 5a$

$\Rightarrow \quad a = 4.9$

Remember $g = 9.8$.

The acceleration of the particle is 4.9 m/s².

b) There is no motion and hence no acceleration perpendicular to the inclined plane so

Shorthand for "applying Newton's second law parallel to the plane with "down" as the positive direction".

Calculation of the component out of the plane of the resultant force.

Mass × component out of the plane of the acceleration.

[N2L, perpendicular, out] $R - 5g \cos 30° = 5 \times 0$

$\Rightarrow \quad R = 5g \cos 30° = 42.435 \ldots$

The reaction between the particle and the plane is 43.4 N (3 s.f.)

In pages 40–42 of chapter 3 a mathematical model for frictional forces was developed. Recall that if a particle is moving across a rough surface, then the frictional force, F, opposes the motion and has magnitude which is taken to be proportional to the normal reaction, R, between the particle and surface; the constant of proportionality, μ, is called the coefficient of friction between the particle and the surface. Thus, for a moving particle, the frictional force is given by $F = \mu R$.

EXAMPLE 5

A particle of mass 20 kg is pulled up a slope inclined at 60° to the horizontal by a force of magnitude P N, parallel to the slope. If the coefficient of friction between the particle and the slope is 0.2, and the acceleration of the particle is 1.8 m/s², find the value of P.

The weight of the particle is 20g N.

Since we are trying to find P it is natural to first apply Newton's second law up the slope:

[N2L, parallel, up] $\quad P - F - 20g \sin 60° = 20 \times 1.8$

$\Rightarrow \quad P = F + 205.740 \ldots$

To find P from this equation we need to know the value of the frictional force, F.
F can be found from the equation $F = \mu R$ once we have found the magnitude of the normal force, R.

R can be found by applying Newton's second law perpendicular to the plane, remembering that there is no acceleration perpendicular to the plane.

[N2L, perpendicular, out] $\quad R - 20g \cos 60° = 20 \times 0$

$\Rightarrow \quad R = 20g \cos 60° = 98$

$F = \mu R \quad \Rightarrow \quad F = 0.2 \times 98 = 19.6$ N

$P = F + 205.740 \ldots \quad \Rightarrow \quad P = 19.6 + 205.740 \ldots = 225.34 \ldots$

$\Rightarrow \quad P = 225$

Newton's Third Law

The third law states that
"*If body A exerts a force on body B then body B exerts an equal and opposite force on body A.*"

In most cases that we will consider the two bodies will be in contact with each other but Newton's third law is true for any two objects regardless of whether they are touching or not. For example, if two magnets are arranged with opposite poles facing each other then the two magnets attract each other:

S	Magnet A	N

S	Magnet B	N

Newton's third law implies that the force of attraction that magnet A experiences due to the presence of magnet B will be equal in magnitude and opposite in direction to the force of attraction that magnet B experiences due to the presence of magnet A.

| S | Magnet A | N | F→ ←F | S | Magnet B | N |

We have already made use of the third law when producing force diagrams to show the forces acting on part of a system.

EXAMPLE 6

A parcel of mass 2 kg rests on the floor of a lift of mass 100 kg which is accelerating upwards at 3 m/s².

a) Draw a force diagram to show the forces acting on the parcel and determine the contact force between the lift floor and the parcel.

b) Draw a force diagram to show the forces acting on the lift and hence find the tension in the lift cable.

c) Draw a force diagram showing the external forces acting on the combined system of lift and parcel. Use this diagram to check your previous answer for the tension.

SOLUTION

Forces on parcel

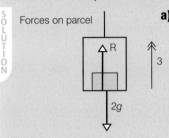

a) The weight of the parcel is $2g$ N.

If R is the normal force between the lift floor and the parcel then applying Newton's second law vertically upwards to the parcel gives

$$R - 2g = 2 \times 3$$
$$\Rightarrow \quad R = 25.6$$

The contact force between the lift floor and the parcel is 25.6 N.

Forces on lift

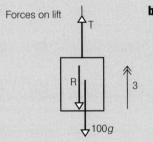

b) The weight of the lift is $100g$ N.

The parcel experiences an upwards force of R from the lift so Newton's third law means that the lift experiences a **downwards** force of R from the parcel.

Applying Newton's second law vertically upwards to the lift gives

$$T - R - 100g = 100 \times 3$$
$$\Rightarrow \quad T = 300 + 25.6 + 980$$
$$\Rightarrow \quad T = 1305.6$$

The tension in the lift cable is 1305.6 N.

Forces on whole system

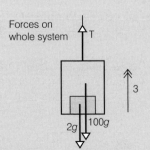

c) When the whole system (lift and parcel together) is considered the only external forces acting are the tension in the cable and the weights of the lift and the parcel.

The force of the lift on the parcel and the force of the parcel on the lift are **not** shown in this diagram since they are internal forces of the whole system and cancel each other out.

Applying Newton's second law vertically upwards to the whole system gives

$$T - 2g - 100g = 102 \times 3$$
$$\Rightarrow \quad T = 1305.6$$

which is in agreement with the previous answer.

EXERCISE 1

1 The diagrams show the forces acting on a particle. In each case, determine whether or not the particle has an acceleration.

a)

b)

2 A particle of mass 3 kg has an acceleration of 4 m/s². Find the magnitude of the resultant force acting on the particle.

3 A particle of mass 4 kg moves in a straight line under the action of a constant force. The particle accelerates from rest to a speed of 6 m/s over a distance of 12 m. Find
a) the acceleration of the particle;
b) the magnitude of the force acting on the particle.

4 A particle of mass 1.5 kg is pulled across a smooth table by a string inclined at 30° to the horizontal. If the acceleration of the particle is 2 m/s² find
a) the tension in the string;
b) the normal force between the table and the particle.

5 A small block of mass 6 kg is pulled across a rough horizontal floor by a horizontal rope. Find the tension in the string if the acceleration of the block is 0.8 m/s² and the coefficient of friction between the block and the floor is 1.2.

6 A block of mass 5 kg is pulled across a rough table by a horizontal string. If the tension in the string is 12 N and the acceleration of the block is 0.8 m/s² find the coefficient of friction between the block and the table.

7 A particle of mass 6 kg is pulled up a smooth plane inclined at 20° to the horizontal by a string parallel to the slope. If the acceleration of the particle is 1.6 m/s² find the tension in the string.

8 A particle of mass 3 kg slides down a rough plane inclined at 35° to the horizontal with acceleration 1.2 m/s². Determine the coefficient of friction between the particle and the plane.

9 A particle of mass 2 kg is pulled up a rough plane inclined at 15° to the horizontal by a string parallel to the slope. If the tension in the string is 18 N and the coefficient of friction between the particle and the plane is 0.4 find the acceleration of the particle.

10 The diagram shows a block of mass 5 kg being pulled across a horizontal surface by a string in which the tension is 30 N. The acceleration of the block is 2 m/s².

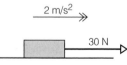

a) Copy the diagram and show all the forces acting on the block on your diagram.
b) Calculate the size of the frictional force between the block and the surface.
c) Calculate the size of the normal reaction between the block and the surface.
d) Calculate the coefficient of friction between the block and the surface.

The string is now moved so that it is inclined at 25° to the horizontal as shown in the diagram.
e) Calculate the new value of the normal reaction between the block and the surface.
f) Calculate the acceleration of the block.

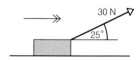

11 A particle of mass 5 kg is pushed up a smooth plane inclined at 30° to the horizontal by a horizontal force of magnitude 40 N.

Find, correct to three significant figures,
a) the acceleration of the particle;
b) the contact force between the particle and the plane.

12 The force of attraction, F N, between the Earth and a body of mass m kg which is x metres from the centre of the Earth is known to be given by an equation of the form

$$F = \frac{km}{x^2}$$

a) Use the fact that a body of mass 5 kg has a weight of 49 N when it is near the Earth's surface and the fact that the radius of the Earth is 6.4×10^6 metres to find the value of the constant k.
b) A spaceship of mass 2000 kg is travelling away from the Earth. At a certain instant, while the rockets are burning, the spaceship is 10 000 km from the centre of the Earth and is accelerating away from the Earth at a rate of 1.5 m/s².

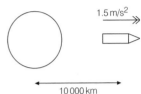

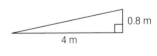

i) Draw a diagram to show the forces acting on the rocket.
ii) Calculate the thrust produced by the rockets.

13 A woman of mass 60 kg stands on the floor of a lift which is accelerating upwards at 1.2 m/s². Find the reaction between the woman and the lift floor.

14 A block of mass 10 kg rests on the floor of a lift. If the reaction between the block and the lift floor is 78 N find the magnitude and direction of the acceleration of the lift.

15 A man of mass 70 kg travels in a lift of mass 200 kg which is accelerating upwards at 1.5 m/s².
a) By drawing a diagram showing the external forces on the whole system of lift + man, determine the tension in the lift cable.
b) By drawing a diagram showing the forces acting on the man, determine the contact force between the man and the lift floor.

16 A car accelerates at a constant rate from 54 km/hr to 72 km/hr in 4 s. If the car has a mass of 750 kg and is working against constant resistances of 1000 N, determine the driving force produced by the car's engine.

17 The diagram shows a rough inclined plane.

A small block of mass 0.8 kg is placed at the top of the plane and when it is released from rest takes 3.5 s to slide down the plane.

Determine the coefficient of friction between the particle and the plane, making clear any assumptions that are made.

Applying Newton's Laws to Systems of Bodies

If we have a system of bodies moving in a straight line with the same acceleration then we can apply Newton's second law either to the whole system or to each of the individual components of the system.

EXAMPLE 7

The diagram shows two sledges, A and B, of masses 20 kg and 25 kg, respectively which are connected by a short horizontal rope. The sledges are pulled across rough horizontal ground by a long rope which is attached to A and makes an angle of 30° with the horizontal.

If the tension in the long rope is 200 N and the coefficient of friction between each sledge and the ground is 0.3, find

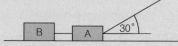

a) the acceleration of the sledges;
b) the tension in the short rope connecting the two sledges.

SOLUTION

a) Start by considering the whole system. The mass of the whole system is 45 kg so the weight of the whole system is $45g$ N.

Applying Newton's second law horizontally and vertically to the whole system:

$$[\text{N2L, } \rightarrow] \qquad 200 \cos 30° - F = 45a \qquad [1]$$

$$[\text{N2L, } \uparrow] \qquad 200 \sin 30° + R - 45g = 0$$
$$\Rightarrow \quad R = 341$$

Since the system is moving, $F = 0.3R = 0.3 \times 341 = 102.3$ so equation [1] can be rewritten as

$$200 \cos 30° - 102.3 = 45a$$
$$\Rightarrow \quad 70.905 \ldots = 45a$$
$$\Rightarrow \quad a = 1.5756 \ldots$$

The acceleration of the sledges is 1.58 m/s² (3 s.f.).

b) To find the tension in the connecting rope we must consider the forces acting on just one of the two sledges.
If T is the tension in the connecting rope then sledge B is being pulled by the force T whilst sledge A is being dragged back by the force T. Concentrating on sledge B, which has a weight of $25g$, gives the force diagram shown.

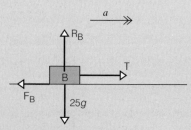

Newton's second law is now applied horizontally and vertically to sledge B:

$$[\text{N2L, } \rightarrow] \qquad T - F_B = 25a \qquad [2]$$

EXAMPLE 7 (continued)

[N2L, ↑] $R_B - 25g = 0$ [3]

$$\Rightarrow \quad R_B = 25g = 245$$

$F_B = 0.3R_B \quad \Rightarrow \quad F_B = 73.5$

Substituting the values for a and F_B into equation [2]

$$T - 73.5 = 25 \times 1.5756 \dots$$

$$\Rightarrow \quad T = 112.891 \dots$$

The tension in the rope connecting the sledges is 113 N (3 s.f.)

When we have a connected system of particles which are moving in different directions – for example two particles connected by a string passing over a pulley – then the accelerations of the particles will be in different directions so it is not possible to apply Newton's second law to the whole system. However, valid equations can be obtained by applying Newton's second law to each of the components of the system separately and the resulting system of equations will enable the motion of the system to be fully described.

EXAMPLE 8

Two particles of mass 6 kg and 9 kg are connected by a light inextensible string passing over a smooth light pulley. Find the acceleration of the system and the tension in the string.

Modelling Assumptions
The wording of this question implies several modelling assumptions that need to be made in order to answer the question:

- a **light** string has no mass – if the string had mass we would need to keep track of how much string was each side of the pulley;
- an **inextensible** string has a constant length which means that, if the string is to remain taut, the velocities (and hence accelerations) of the particles at each end of the string must have the same magnitude components in the direction of the string;
- a **light** pulley has no mass so the rotational motion of the pulley can be ignored;
- a **smooth** pulley ensures that the tension in the string is the same both sides of the pulley.

The 6 kg mass has weight $6g$ N and the 9 kg mass has weight $9g$ N.
A diagram could be drawn to show the motion of the system and clear force diagrams showing the forces acting on each particle.

The magnitude of the upward acceleration of the 6 kg block is the same as the magnitude of the downward acceleration of the 9 kg block since the string is inextensible.

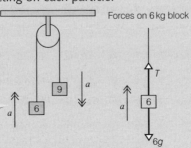

Forces on 6 kg block   Forces on 9 kg block

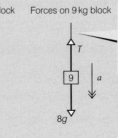

The tension in each portion of the string is the same since the pulley is smooth.

EXAMPLE 8 (continued)

We could combine the diagrams into a single diagram showing the acceleration of each particle and all the forces acting on each particle.

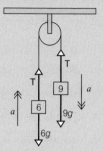

We cannot apply Newton's second law to the whole system since the whole system is not moving in a single direction.

We can apply Newton's second law to the 6 kg particle and the 9 kg particle **separately**.

[N2L, ↑, 6 kg block] $T - 6g = 6a$ [1]

[N2L, ↓, 9 kg block] $9g - T = 9a$ [2]

Newton's second law has given a pair of simultaneous equations for the unknown values T and a. These can be quickly solved by simply adding the two equations which eliminates T:

[1] + [2] $(T - 6g) + (9g - T) = 6a + 9a$

$\Rightarrow \quad 3g = 15a$

$\Rightarrow \quad a = \dfrac{1}{5}g = 1.96$

The acceleration of the system is 1.96 m/s^2.

Substituting the value of a into equation [1] gives

$T - 58.8 = 6 \times 1.96$

$\Rightarrow \quad T = 70.56$

The tension in the string is 70.56 N.

> Alternatively, rewrite equations [1] and [2] as
>
> $T - 6a = 58.8$
>
> and
>
> $T + 9a = 88.2$
>
> and then use an algebraic method or your graphical calculator to solve the equations.

EXAMPLE 9

A particle P of mass 3 kg rests on a smooth plane inclined at 30° to the horizontal. It is connected by a light inextensible string passing over a smooth light pulley at the top of the plane, to a particle of mass 5 kg which is on a smooth plane inclined at 60° to the horizontal.

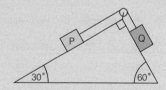

Find the acceleration of the system and the tension in the string, giving your answers correct to three significant figures.

Since Q is heavier than P and on the steeper slope, Q will accelerate down the slope and P will accelerate up the slope. The force diagrams for each particle are shown below:

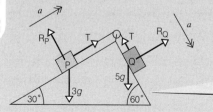

> This diagram combines the force diagram for the 3 kg particle with the force diagram for the 5 kg particle. Remember that it is **not** possible to apply Newton's second law to the whole system since the two particles are moving in different directions.

EXAMPLE 9 (continued)

Applying Newton's second law to each particle in the direction that it is accelerating gives

$$T - 3g \sin 30° = 3a \qquad [1]$$

for P, and

$$5g \sin 60° - T = 5a \qquad [2]$$

for Q.

Adding the two equations will eliminate T:

$$[1] + [2] \qquad 5g \sin 60° - 3g \sin 30° = 8a$$
$$\Rightarrow \qquad 27.7352 \ldots = 8a$$
$$\Rightarrow \qquad a = 3.46690 \ldots$$

and substituting this value into equation [1] gives

$$T = 3g \sin 30° + 3 \times 3.46690 \ldots = 25.1007 \ldots$$

So, correct to three significant figures, the acceleration of the system is 3.47 m/s^2 and the tension in the string is 25.1 N.

EXAMPLE 10

A particle of mass 2 kg rests on a rough plane inclined at 40° to the horizontal. It is connected by a light inextensible string, passing over a smooth light pulley at the top of the plane, to a block of mass 2.5 kg which is hanging freely. If the coefficient of friction between the particle and the inclined plane is 0.4, find

a) the acceleration of the system;
b) the tension in the string;
c) the magnitude and direction of the force in the pulley support.

S
O
L
U
T
I
O
N

The diagram shows the forces acting on each of the two particles.
Consider first the 2 kg particle.

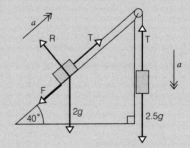

[N2L, parallel to slope, up]
$$T - F - 2g \sin 40° = 2a$$
$$\Rightarrow \qquad T - F - 12.5986 \ldots = 2a \qquad [1]$$

[N2L, perpendicular to slope, out]
$$R - 2g \cos 40° = 0$$
$$\Rightarrow \qquad R = 2g \cos 40° = 15.0144 \ldots$$

Since the 2 kg mass is moving,
$$F = \mu R = 0.4 \times 15.0144 \ldots = 6.0057 \ldots$$
and substituting this value into equation [1] gives

$$T - 6.0057 \ldots - 12.5986 \ldots = 2a$$
$$\Rightarrow \qquad T - 18.6044 \ldots = 2a \qquad [2]$$

EXAMPLE 10 (continued)

Now consider the 2.5 kg particle.

[N2L, ↓] $2.5g - T = 2.5a$

$\Rightarrow \quad 24.5 - T = 2.5a$ [3]

Adding equations [2] and [3] will eliminate T.

[2] + [3] $(T - 18.6004 \dots) + (24.5 - T) = 2a + 2.5a$

$\Rightarrow \quad 5.8955 \dots = 4.5a$

$\Rightarrow \quad a = 1.3101 \dots$

and substituting this into [2] gives $T = 18.6044 \dots + 2 \times 1.3101 \dots = 21.2246 \dots$

So, correct to three significant figures, the acceleration of the particles is 1.31 m/s^2 and the tension in the string is 21.2 N.

Finally, consider the forces acting on the pulley.

In applying Newton's second law to the pulley, we are assuming the pulley is light so has negligible mass and can therefore write

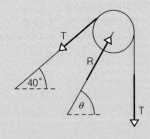

[→] $R \cos \theta - T \cos 40° = 0$

$\Rightarrow \quad R \cos \theta = 16.259 \dots$

[↑] $R \sin \theta - T - T \sin 40° = 0$

$\Rightarrow \quad R \sin \theta = 34.867 \dots$

These two results give the horizontal and vertical components of R so trigonometry can be used to find R and θ:

$$R^2 = 16.259\dots^2 + 34.867\dots^2 = 1480.10 \dots$$

$\Rightarrow \quad R = 38.47 \dots$

$$\tan \theta = \frac{34.867 \dots}{16.259 \dots}$$

$\Rightarrow \quad \theta = 64.999 \dots$

Correct to 3 significant figures, the force in the pulley support has magnitude 38.5 N and makes an angle of 65.0° with the horizontal.

EXERCISE 2

1 (1 tonne = 1000 kg)

A tug of mass 120 tonnes is pulling a barge of mass 480 tonnes. Both the tug and the barge experience a resistance of R N per tonne. When the engines produce a thrust of 14.4 kN the tug and the barge accelerate at 0.01 m/s^2.

a) Draw a diagram to show the forces acting on the whole system of barge and tug together.

b) Find the value of R.

c) Draw a diagram to show the forces acting on just the tug. Hence find the tension in the tow-rope.

2 The diagram shows two blocks, A and B, of mass 5 kg and 2 kg, respectively which are linked together by a rod and are moving across a smooth horizontal table. The block A is pulled by a horizontal force of 28 N.

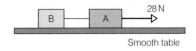

a) Draw a diagram to show all the forces acting on the block A. Draw a separate diagram to show all the forces acting on the block B.

b) Write down an equation relating the forces acting on the block A to its acceleration. Write down another equation relating the forces acting on block B to its acceleration.

c) Hence find the acceleration of the blocks and the size of the force in the connecting rod. This situation is used to model a child's train and carriage toy set being pulled by a string across a table. What assumptions have been made in this model that might affect the actual motion of the toy?

3 A car of mass 800 kg moves along a horizontal road pulling a trailer of mass 200 kg. If the car and trailer are accelerating at 0.6 m/s² when the resistance to the car's motion is 750 N and the resistance to the trailer's motion is 250 N, find

a) the driving force produced by the car's engine;

b) the force in the tow-bar linking the car to the trailer.

4 Two particles of mass 2.3 kg and 2.6 kg are connected by a light inextensible string which passes over a smooth fixed pulley. Find the acceleration of the system and the tension in the string.

5 Two particles of mass m kg and 1.5 kg are connected by a light inextensible string which passes over a smooth fixed pulley. If the 1.5 kg particle has a downward acceleration of 0.7 m/s² find the value of m and the tension in the string.

6 Two small blocks, A and B, of mass 3.4 kg and 3.6 kg, respectively are connected by a string which passes over a small peg. Initially A is held at rest 0.8 m below the peg and B is below A.

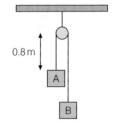

If the system is released from rest, find the acceleration of A and the time that elapses before it hits the peg.

State three assumptions that have been made in answering this question.

7 A particle of mass 4.8 kg rests on a smooth horizontal table. It is connected by a light inextensible string passing over a smooth pulley at the edge of the table to a particle of mass 0.8 kg which is hanging freely.

Find the acceleration of the system and the tension in the string.

8 The diagram shows a block of mass 8 kg moving on a horizontal table. The block is connected by means of a string passing over a small pulley to a particle of mass 2 kg. The particle is accelerating downwards at 1.5 m/s².

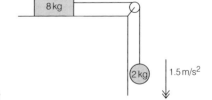

a) Calculate the tension in the string.

b) Calculate the magnitude of the frictional force between the 8 kg block and the table.

c) Calculate the coefficient of friction between the 8 kg block and the table.

Write down two simplifications or assumptions that have been made in answering this question.

9 A block of mass 1.4 kg rests on a smooth plane inclined at 30° to the horizontal. The particle is connected by a light inelastic string passing over a smooth pulley at the top of the plane to a particle of mass 2.1 kg which is hanging freely. Find the acceleration of the system and the tension in the string.

10 A block of mass 5 kg rests on a rough plane inclined at 38° to the horizontal. The coefficient of friction between the block and the plane is 0.2. The particle is connected by a light inelastic string passing over a smooth pulley at the top of the plane to a particle of mass 2 kg which is hanging freely. Find the acceleration of the system and the tension in the string.

11 In the diagram, A and B are particles of mass 1.3 kg and 1.5 kg, respectively. Initially they are held at rest on the slopes PQ and QR as shown. They are connected by a light inextensible string passing over a smooth fixed light pulley at Q.

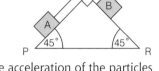

a) If the planes PQ and QR are both smooth, prove that the acceleration of the particles is $\dfrac{7\sqrt{2}}{20}$ m/s² and find the exact value of the tension in the string.

b) If the plane QR is smooth but the plane PQ is rough and the coefficient of friction between PQ and A is 0.1 find, correct to three significant figures, the acceleration of the system.

12 A particle A of mass 6.5 kg is connected by a light inextensible string passing over a smooth fixed light pulley to a scale pan, C, of mass 0.5 kg which contains a block B of mass 7 kg. Find

a) the acceleration of the system;
b) the tension in the string;
c) the reaction between B and C.

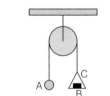

13 Two blocks, P and Q, of mass 4 kg and 3 kg, respectively, are placed on a rough slope inclined at 15° to the horizontal with Q above P. The blocks are linked by a light rod, parallel to the slope. Q is pulled up the slope by a string parallel to the slope.

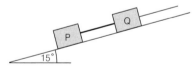

The coefficient of friction between P and the slope is 0.2 and the coefficient of friction between Q and the slope is 0.4.
If the acceleration of the system is 0.5 m/s²:

a) by considering the forces acting on P, determine the force in the connecting rod;
b) determine the tension in the string.

14 The diagram shows three objects connected by strings which pass over pulleys.

The object A, which has mass 6m, has downward acceleration of $\dfrac{g}{10}$.

a) Find the tension in the string connecting A and B.
b) By considering the motion of B, find the tension in the string connecting B and C.
c) Find the coefficient of friction between C and the horizontal surface it is moving across.
d) What assumptions have you made in answering this question?

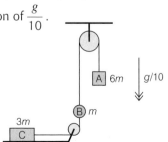

Having studied this chapter you should know

- that if the resultant force acting on a particle is zero then the particle is at rest or moving with constant velocity and if the resultant force is non-zero then the particle is accelerating (Newton's first law)

- that if a non-zero resultant force acts on a particle then the direction of the resultant force is the same as the direction of the acceleration and the magnitude of the resultant force equals the product of the mass and the magnitude of the acceleration (Newton's second law)

- that Newton's second law is frequently applied as
  Component of the resultant force in a given direction
      = mass × component of the acceleration in the given direction

- that if body A exerts a force on body B then body B exerts an equal and opposite force on body A (Newton's third law)

- how to use Newton's laws to model situations as the motion of two or more connected particles

REVISION EXERCISE

1 The diagram shows a block of mass 4 kg being pulled across a rough floor by a string.

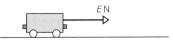

a) Copy the diagram and show all the forces acting on the block.
b) Calculate the size of the reaction force between the floor and the block.

The tension in the string is 20 N and the acceleration of the block is 2 m/s^2.
c) Calculate the magnitude of the frictional force between the block and the floor.
d) Calculate the coefficient of friction between the block and the floor.

2 An engine pulls a truck of mass 6000 kg along a straight horizontal track, exerting a constant horizontal force of magnitude E N on the truck. The resistance to motion of the truck has magnitude 400 N, and the acceleration of the truck is 0.2 m/s^2. Find the value of E.

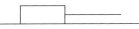

E N

(OCR May 2002 M1)

3 The diagram shows a package of mass 15 kg resting on the floor of a lift.
a) Find the size of the contact force between the package and the floor of the lift if the lift is accelerating upwards at 3 m/s^2.
b) Find the size and direction of the lift's acceleration if the contact force between the package and the floor of the lift is 130 N.

4 When a railway engine of mass 20 tonnes travelling at a speed of 20 km/hr runs into the buffers at the end of the railway track the buffers move a distance of 3.5 metres before the engine comes to rest.
Estimate
a) the deceleration of the engine;
b) the force exerted by the buffers on the engine.

What assumptions have you made in answering this question?

5 The diagram shows two blocks, of mass $5m$ and $2m$, connected by a string passing over a small pulley. The $2m$ mass hangs freely while the $5m$ mass is on a rough table.

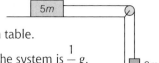

When the system is released from rest the acceleration of the system is $\frac{1}{5}g$.

a) Draw a diagram showing the forces acting on the $2m$ block and hence find an expression in terms of m and g for the tension in the string.

b) Determine, in terms of m and g, the size of the frictional force between the $5m$ block and the table and hence show that the coefficient of friction between this block and the table is $\frac{3}{25}$.

State two assumptions that you have made in answering this question.

6 A sledge of mass 25 kg is on a plane inclined at $30°$ to the horizontal. The coefficient of friction between the sledge and the plane is 0.2.

a) The sledge is pulled up the plane, with constant acceleration, by means of a light cable which is parallel to the line of greatest slope. The sledge starts from rest and acquires a speed of 0.8 m/s after being pulled for 10 seconds. Ignoring air resistance, find the tension in the cable.

b) On a subsequent occasion the cable is not used and two people of total mass 150 kg are seated on the sledge. The sledge is held at rest by a horizontal force of magnitude P N. Find the least value of P which will prevent the sledge from sliding down the plane.

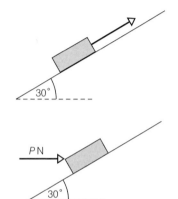

(OCR May 2002 M1)

7 Find the acceleration of the system and the tension in the string:

a) if all contacts are smooth;

b) if the coefficient of friction between the blocks and the planes is 0.1.

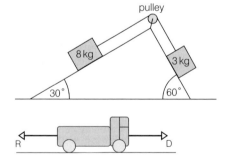

8 A 2000 kg motor van is travelling along a horizontal road with its engine working at a constant rate.

The motion of the van can be modelled by assuming that the van is moving under the action of two forces – a driving force, D, and a resistive force, R. Each of these forces varies with the speed, v, of the van as shown in the table below.

v (m/s)	10	20	30	40
D (N)	4500	2250	1500	1125
R (N)	70	280	630	1120

a) Find a rule linking D to v. **b)** Find a rule linking R to v.

c) Find the acceleration of the van at the instant when its speed is 25 m/s.

d) Find the maximum possible speed of the van.

9 The diagram shows a block of mass 8 kg being pulled across a smooth surface.

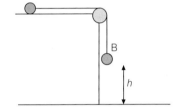

   a) Copy the diagram and show all the forces acting on the block.
   b) Calculate the acceleration of the block.
   c) Calculate the magnitude of the normal reaction between the block and the surface.

This model is to be used to describe a toboggan being pulled across a horizontal snow-covered surface. Write down a simplification which has been made in this model. How is this simplification likely to affect your answers to (b) and (c)?

10 A particle P is released from rest at the top of a smooth plane of length 5 m, which is inclined at 5° to the horizontal. Neglecting air resistance, find
   a) the acceleration of A down the plane;
   b) the time taken for P to reach the bottom of the plane;
   c) the speed with which P reaches the bottom of the plane.

   (OCR Jun 2001 M1)

11 Particles A and B, each of mass 0.6 kg, are joined by a light inextensible string. The string passes over a smooth pulley at the edge of a smooth horizontal platform. A is held at rest on the platform. B hangs vertically below the pulley at a height h above the floor. A is released, with the string taut, and the particles start to move. There is no air resistance.

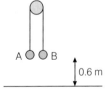

   a) Find the tension in the string and the acceleration.
   b) Hence find the speed of A after it has travelled a distance of 2 m.
   c) When A has moved a distance of 2 m it becomes detached from the string. From this instant B takes a further 0.2 s to reach the floor. Find the value of h.
   d) Find also the total time for which B is in motion before it reaches the floor.

   (OCR Jan 2001 M1)

12 Two particles A and B, of mass 4 kg and 3 kg, respectively, are attached to the ends of a string of length 2.4 m which passes over a small peg fixed at a height of 1.8 m above a table.

The system is released from rest with each particle 0.6 m above the table.

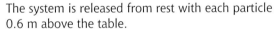

During the subsequent motion the acceleration of A is a m/s^2 and the tension in the string is T N.
   a) By considering the motion of A obtain an equation relating T and a.
   b) Obtain another equation relating T and a and hence show that $a = 1.4$ and find the value of T.
   c) Find the speed with which A hits the table.
   d) Write down two simplifications that you have made in modelling this situation and which are likely to affect the final answers.

97

13 A bullet of mass 60 grams enters a block of wood 5 cm thick with a velocity of 120 m/s and leaves the block with a velocity of 75 m/s.
a) Calculate the retardation of the bullet whilst it is passing through the block.
b) Calculate the time that the bullet takes to pass through the block.
c) With what size force does the block resist the bullet's motion?

What assumptions have you made in answering this question?

14 The diagram shows a particle being pulled up a smooth slope by means of a rope.
a) If the particle has mass 70 kg and the tension in the rope is 300 N, calculate the acceleration of the particle up the slope.

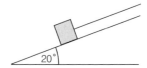

b) This model is used to analyse the motion of a skier being pulled up a ski-slope in Switzerland by a drag lift. Write down one simplification which has been made in this model and which is likely to affect the actual motion of the skier.
c) Give a reason why the acceleration might be slightly different for a skier on a mountain in Canada.

15 The diagram shows a parcel of mass 60 kg sitting on a trolley of mass 15 kg.

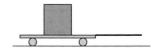

The trolley is pulled by a horizontal rope and the tension in this rope is T N.
There are no resistive forces between the wheels of the trolley and the floor.
The trolley is accelerating to the right at 0.4 m/s² and the parcel is at rest relative to the trolley.

The normal reaction between the trolley and the parcel is R N and the frictional force between the trolley and the parcel is F N.
a) **i)** Show in a diagram the three forces acting on the parcel.
ii) Show in a second diagram the forces acting on the trolley.

b) Calculate the values of R, F and T.

16 The diagram shows two particles P and Q, of mass 6 kg and 4 kg, respectively, connected to each other by a string passing over a smooth pulley at C. The particle P rests on a rough horizontal surface AB while Q hangs vertically below C and is initially 3 m above ground level. The system is released from rest in the position shown. After 2 seconds, Q hits the ground.

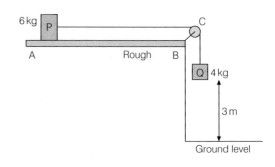

a) Determine the acceleration of the system.
b) Determine the speed with which Q hits the ground.
c) By considering the forces acting on Q, determine the tension in the string.
d) Find the frictional force between P and the table.
e) State two assumptions that you have made in answering this question.

If the experiment was repeated on the Moon rather than on the Earth describe, with reasons, what would happen. (The acceleration due to gravity on the Moon is roughly one sixth of the figure on the Earth.)

6 Linear Momentum and Collisions

The purpose of this chapter is to enable you to

- find the momentum of a particle moving in a straight line
- use the conservation of momentum when considering collisions between particles

Momentum and Impulse

In the previous chapter we have seen how a detailed knowledge of the forces acting on a particle enables the acceleration and hence the motion of the particle to be deduced through an application of Newton's second law.

There are, though, situations such as collisions between railway trucks or balls being hit by bats where unknown forces of considerable size may be acting for periods of time that may be very short and/or very difficult to measure. In these situations it is impossible to directly apply Newton's second law.

Consider now two particles, A and B, whose masses are m_A and m_B, respectively, and which are moving along the same straight line in such a way that they are going to collide. Suppose that the velocities just before the collision are u_A and u_B.

When the two particles collide there will be sudden changes in the motion of each of the particles caused by the contact forces between the two particles. Remember that Newton's third law means the force that B exerts on A at any time is equal and opposite to the force that A exerts on B at that time.

After a short time, once the collision phase is over, the two trucks are moving with their new velocities.

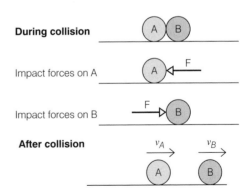

If we assume that the contact force between the two particles during the collision phase is constant and that this phase lasts t seconds then an important result linking m_A, m_B, u_A, u_B, v_A and v_B may be obtained from applying Newton's second law and the constant acceleration equations to each of the particles separately.

Consider first what is happening to B during the collision. Let a_B be the acceleration of B during the collision phase.

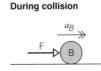

Newton's second law, applied to the right, gives

$$F = m_B a_B \qquad\qquad [1]$$

and the constant acceleration equation "$v = u + at$" applied to the t seconds of the collision gives

$$v_B = u_B + a_B t$$

$$\Rightarrow \quad a_B = \frac{v_B - u_B}{t} \qquad\qquad [2]$$

Substituting this expression for a_B into equation [1] gives

$$F = m_B \frac{v_B - u_B}{t}$$

$$\Rightarrow \quad Ft = m_B v_B - m_B u_B \qquad\qquad [3]$$

Before we consider what is happening to particle A during the collision it is worth looking at equation [3] in some detail.

The term Ft gives the **impulse** of a constant force F acting for time t. Since t is a scalar quantity and F is a vector quantity, impulse is a vector quantity and the direction of the impulse is the same as the direction of the force. Since impulse is the product of force and time it is measured in units of Ns.

> Recall that if k is a positive scalar quantity and p is a vector then kp is a vector in the same direction as p and k times the size of p.

The right-hand side of equation [3] has two terms each of which is a product of a mass with a velocity. For a particle of mass m kg moving with velocity v m/s, the quantity mv is called the **momentum of the particle**. Momentum is a vector quantity and has the same direction as the velocity. Since momentum is the product of mass and velocity, the units of momentum can be written as kg m/s.

Equation [3],

$$Ft = m_B v_B - m_B u_B$$

can be interpreted verbally as

impulse of the contact force of A on B
    = final momentum of B – initial momentum of B

or

impulse of the contact force of A on B
    = change in momentum of B during the collision.

This means that the units of momentum must be the same as the units of impulse. We can therefore express the units of momentum as either kg m/s or Ns.

> If you go on to study the M2 module you will study impulses in much more detail.

Now consider what is happening to A during the collision and let a_A be the acceleration of A during the collision phase.

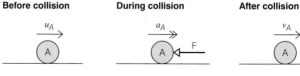

Newton's second law, applied to the right, gives

$$-F = m_A a_A \qquad \text{[4]}$$

and the constant acceleration equation "$v = u + at$" applied to the t seconds of the collision gives

$$v_A = u_A + a_A t$$
$$\Rightarrow \quad a_A = \frac{v_A - u_A}{t} \qquad \text{[5]}$$

Substituting this expression for a_A into equation [1] gives

$$-F = m_A \frac{v_A - u_A}{t}$$
$$\Rightarrow \quad -Ft = m_A v_A - m_A u_A \qquad \text{[6]}$$

Now, adding equations [3] and [6] gives

$$Ft + (-Ft) = (m_B v_B - m_B u_B) + (m_A v_A - m_A u_A)$$
$$\Rightarrow \quad 0 = m_A v_A + m_B v_B - m_A u_A - m_B u_B$$
$$\Rightarrow \quad m_A u_A + m_B u_B = m_A v_A + m_B v_B \qquad \text{[7]}$$

The expression $m_A u_A + m_B u_B$ gives the total momentum of the two particles just **before** the collision whilst the expression $m_A v_A + m_B v_B$ gives the total momentum of the two particles just **after** the collision.

Equation [7] states that the total momentum just before the collision is the same as the total momentum just after the collision. This important principle is known as the **conservation of momentum**.

The derivation of equation [7] presented here has assumed that the contact force was constant during the collision phase and that the impulses from the contact forces were the only impulses acting during the collision phase.

The assumption of constant contact forces during the collision phase is unnecessary, but a derivation of the conservation of momentum for non-constant contact forces is rather more complex.

However, the assumption that there are no other impulses acting during the collision phase is necessary: if particle B received a sudden kick (or impulse) to the left as it was colliding with particle A then the kick would drastically affect the motion of both particles. The other forces acting on the particles must also contribute no significant impulse – this is usually the case since the time involved in most collisions is quite small and the other forces acting are modest in size when compared to the contact force present during the collision.

Conservation of Momentum for Collisions
If two particles collide in such a way that no external impulses act then the total momentum of the system immediately before the collision equals the total momentum of the system immediately after the collision.

Applying the Conservation of Momentum to Collisions

Separate "before collision" and "after collision" diagrams should be clearly drawn when considering collisions. Make sure that the directions of the velocities are clear in these diagrams.

Remember that the conservation of momentum equation is a **vector equation** so, in one-dimensional problems, it must be clear which direction is being considered the positive direction.

EXAMPLE 1

A truck of mass 1100 kg moving with speed 5 m/s collides with a truck of mass 1400 kg which is moving in the same direction as the first truck at 2 m/s. If the trucks automatically couple when they collide, what is their speed after the collision?

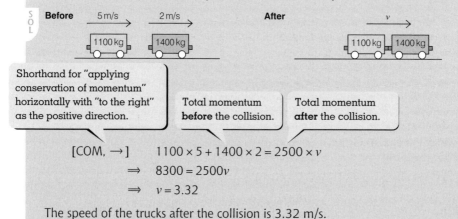

Shorthand for "applying conservation of momentum" horizontally with "to the right" as the positive direction.

Total momentum **before** the collision.

Total momentum **after** the collision.

$$[\text{COM}, \rightarrow] \quad 1100 \times 5 + 1400 \times 2 = 2500 \times v$$
$$\Rightarrow \quad 8300 = 2500v$$
$$\Rightarrow \quad v = 3.32$$

The speed of the trucks after the collision is 3.32 m/s.

EXAMPLE 2

P and Q are two particles of mass 5 kg and m kg, respectively. They collide head on when their velocities are 2.8 m/s and 2 m/s in opposite directions. After the collision the velocity of P is 2 m/s in the same direction as before the collision but the velocity of Q is 3 m/s in the opposite direction to its previous motion. Determine the value of m.

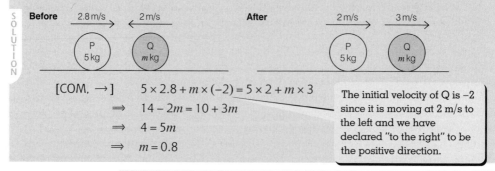

$$[\text{COM}, \rightarrow] \quad 5 \times 2.8 + m \times (-2) = 5 \times 2 + m \times 3$$
$$\Rightarrow \quad 14 - 2m = 10 + 3m$$
$$\Rightarrow \quad 4 = 5m$$
$$\Rightarrow \quad m = 0.8$$

The initial velocity of Q is −2 since it is moving at 2 m/s to the left and we have declared "to the right" to be the positive direction.

EXAMPLE 3

Two particles, S and T, of mass $2m$ and $3m$, respectively, collide head on when their velocities are $6u$ and ku in opposite directions. If they coalesce on impact find the magnitude and direction of the final velocity of the particles, taking care to consider all possible values of the constant k.

Coalesce means join up and become a single body.

EXAMPLE 3 (continued)

Before $6u$ ku **After** v

S
Mass = $2m$

T
Mass = $3m$

In this case we can't be sure whether, after the collision, the combined mass will move to the right or to the left.

When drawing the diagram, this doesn't matter as long as the Conservation of Momentum equation is used accurately. If v is positive then the direction given in the diagram is correct but if v is negative then the direction of motion is opposite to that given in the diagram.

$$[COM, \rightarrow] \quad 2m \times 6u + 3m \times (-ku) = 5m \times v$$
$$\Rightarrow \quad 12mu - 3mku = 5mv$$
$$\Rightarrow \quad 5mv = 3mu(4 - k)$$
$$\Rightarrow \quad v = \frac{3}{5}(4 - k)u$$

If $k < 4$ then v is positive so the combined particle moves in the same direction as S's original motion with speed $\frac{3}{5}(4 - k)u$.

If $k = 4$ then $v = 0$ and the combined particle is stationary after the collision.

If $k > 4$ then v is negative which means that the combined particle moves in the same direction as T's original motion with speed $\frac{3}{5}(k - 4)u$.

EXAMPLE 4

A truck, C, of mass 1500 kg moving at 6 m/s collides head on with a stationary truck, D. After the collision the speed of D is 3 m/s and the speed of C is 2 m/s. Determine the possible values of the mass of truck D.

Suppose that truck D has mass m kg.

 Before 6 m/s   **After** 2 m/s   3 m/s

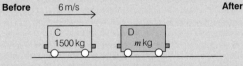

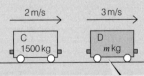

C
1500 kg

D
m kg

C
1500 kg

D
m kg

$$[COM, \rightarrow] \quad 1500 \times 6 + m \times 0 = 1500 \times 2 + m \times 3$$
$$\Rightarrow \quad 9000 = 3000 + 3m$$
$$\Rightarrow \quad 3m = 6000$$
$$\Rightarrow \quad m = 2000$$

The total momentum **before** the collision is to the right; the total momentum **after** the collision must also be to the right so at least one of the trucks must move to the right.

C moving to the right and D moving to the left is impossible so we can be certain that **D moves to the right**.

C could move to the left or the right. Assume to begin with that it moves to the right.

EXAMPLE 4 (continued)

Now suppose that after the collision truck C moves **to the left** with speed 2 m/s:

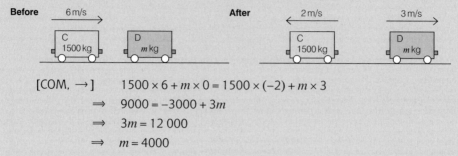

$$[\text{COM}, \rightarrow] \qquad 1500 \times 6 + m \times 0 = 1500 \times (-2) + m \times 3$$
$$\Rightarrow \quad 9000 = -3000 + 3m$$
$$\Rightarrow \quad 3m = 12\,000$$
$$\Rightarrow \quad m = 4000$$

The mass of truck D could be 2000 kg or 4000 kg.

EXAMPLE 5

T and F are two points at the top and bottom, respectively, of a 50 m high tower with T directly above F. At the same instant, an object, A, of mass 0.2 kg is released from rest at T and another object B, of mass 0.4 kg is thrown vertically upwards from F with an initial speed of 20 m/s. When the particles collide they coalesce to form a single particle.

a) Determine the time that elapses before A and B collide and determine the position of the particles when they collide.
b) Find the velocity of each particle just before the collision.
c) Find the velocity of the combined particle just after the collision.
d) Find the time that elapses from the start of the motion to the moment when the combined particle reaches F.
e) On a single diagram, sketch the velocity–time graphs for A and B and the combined particle.
f) State two assumptions made in answering this question.

a)

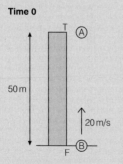

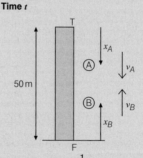

Using the constant acceleration equation $x = ut + \dfrac{1}{2}at^2$, at time t object A will have fallen a distance x_A where

$$x_A = 0 \times t + \frac{1}{2} \times 9.8 \times t^2 = 4.9t^2$$

and object B will be at height x_B above F where

$$x_B = 20 \times t + \frac{1}{2} \times (-9.8) \times t^2 = 20t - 4.9t^2$$

Modelling Assumption
We are assuming there is no air resistance so each particle experiences a constant downward acceleration of 9.8 m/s².

EXAMPLE 5 (continued)

The particles collide when

> **Modelling Assumption**
> This equation certainly assumes the two objects are particles.

$$x_A + x_B = 50$$
$$\Rightarrow \quad 4.9t^2 + 20t - 4.9t^2 = 50$$
$$\Rightarrow \quad 20t = 50$$
$$\Rightarrow \quad t = 2.5$$

The two particles collide 2.5 s after the start of the motion. At this instant

$$x_B = 20 \times 2.5 - 4.9 \times (2.5)^2 = 19.375$$

so the objects collide at a point 19.375 m above F.

b) Using the constant acceleration equation $v = u + at$ with $t = 2.5$ gives

$$v_A = 0 + 9.8 \times 2.5 = 24.5$$

and

$$v_B = 20 + (-9.8) \times 2.5 = -4.5$$

> For B's motion, we have been treating "up" as the positive direction so a velocity of −4.5 means B is moving **down** at 4.5 m/s.

So, just before the collision, A is moving down at 24.5 m/s and B is moving down at 4.5 m/s.

c) Before

0.2 kg (A) ↓ 24.5

0.4 kg (B) ↓ 4.5

After

0.6 kg (A B) ↓ w

> Remember "before" and "after" diagrams are an essential part of the analysis of any collision!

$$[\text{COM, } \downarrow] \quad 0.2 \times 24.5 + 0.4 \times 4.5 = 0.6 \times w$$
$$\Rightarrow \quad 0.6w = 6.7$$
$$\Rightarrow \quad w = 11.166666 \ldots$$

Immediately after the collision, the velocity of the combined object is 11.2 m/s (3 s.f.)

$$\Rightarrow \quad w = 11.2 \text{ m/s} \qquad \text{to 3 s.f.}$$

d) The combined particle is now in free fall with an initial downward velocity of 11.166 ... m/s. The constant acceleration equation $x = ut + \frac{1}{2}at^2$ can be used to determine how long it takes to fall 19.375 m:

$$19.375 = (11.166 \ldots)t + \frac{1}{2} \times 9.8 \times t^2$$

$$\Rightarrow \quad 4.9t^2 + (11.166 \ldots)t - 19.375 = 0$$
$$\Rightarrow \quad t = 1.152 \ldots$$

Total time for motion = 2.5 + 1.152 ... = 3.652 ...

So the total time that elapses from the start of the motion until the combined particle reaches F is 3.65 s (3 s.f.).

EXAMPLE 5 (continued)

e) Treating "down" as the positive direction for all the velocities, the velocity–time graph for the problem is shown below.

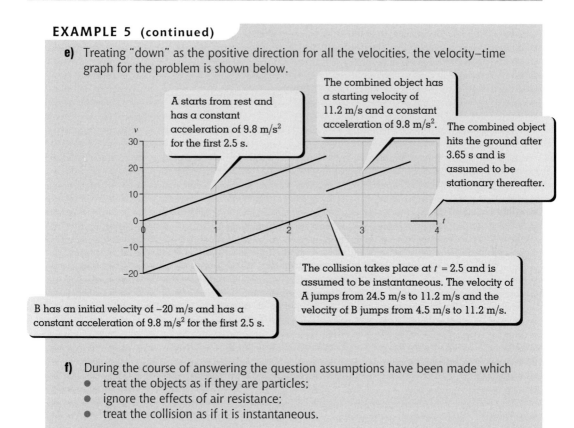

A starts from rest and has a constant acceleration of 9.8 m/s² for the first 2.5 s.

The combined object has a starting velocity of 11.2 m/s and a constant acceleration of 9.8 m/s².

The combined object hits the ground after 3.65 s and is assumed to be stationary thereafter.

The collision takes place at $t = 2.5$ and is assumed to be instantaneous. The velocity of A jumps from 24.5 m/s to 11.2 m/s and the velocity of B jumps from 4.5 m/s to 11.2 m/s.

B has an initial velocity of –20 m/s and has a constant acceleration of 9.8 m/s² for the first 2.5 s.

f) During the course of answering the question assumptions have been made which
- treat the objects as if they are particles;
- ignore the effects of air resistance;
- treat the collision as if it is instantaneous.

EXERCISE 1

1 a) A particle moving with speed 8 m/s has momentum of magnitude 40 Ns. What is the mass of the particle?

b) A particle of mass 0.008 kg has momentum of magnitude 2 Ns. What is the speed of the particle?

c) Find the magnitude of the momentum of a lorry of mass 40 tonnes travelling at 108 km/hr.

2 A stone of mass 0.3 kg is dropped from rest from a tower of height 20 m. Calculate the magnitude of the momentum of the stone just before it hits the ground.

3 A truck of mass 1000 kg approaches at 8 m/s a second truck of mass 600 kg which is initially stationary. After the impact the second truck moves with speed 6 m/s. What is the speed of the first truck after the impact?

4 A bullet of mass 0.05 kg is fired into a block of wood of mass 2 kg which is at rest on a smooth table, with speed 200 m/s. Assuming that the bullet stays embedded in the block, find the speed of the block after the impact.

5 A car of mass 800 kg travelling at 90 km/hr has a head-on crash with a van of mass 2400 kg travelling at 60 km/hr in the opposite direction to the car. Assuming that after the impact the car and van are locked together and the coefficient of friction between the road and the crash debris is 0.4, find
   a) the speed in m/s of the crash debris after the impact;
   b) the distance that the crash debris will travel before coming to rest.

6 Immediately before a collision, particles A and B, of masses 4 kg and 6 kg, respectively, are moving in opposite directions with speeds of 10 m/s and v m/s, respectively. During the collision, the direction of B's motion is reversed and B's speed immediately after the collision is 5 m/s. Immediately after the collision, A is moving with speed 2 m/s. Calculate the possible values of v.

7 Two particles, A and B, each of mass m moving on a smooth horizontal table in opposite directions with speeds $6u$ and $4u$, respectively, collide directly. As a result the velocity of B is reversed in direction but remains unchanged in magnitude.
   a) Find the velocity of A after the collision.

The particle B continues to move at speed $4u$ until it catches up with and coalesces with a particle C, of mass $2m$, which is moving along the same line with speed $2u$.
   b) Find the velocity of the combined particle after the impact.

8 A bullet of mass 0.06 kg travelling horizontally at 100 m/s hits a stationary piece of wood, passes through it and emerges with a horizontal speed of 40 m/s. If the block is then moving with speed 0.3 m/s, find the mass of the block of wood.

9 A truck of mass 4000 kg travelling at 15 m/s along a railway line catches up with a second truck which is moving along the same railway line at 6 m/s in the same direction as the first truck.
After the collision, the speed of the second truck is 10 m/s and the speed of the first truck is 3 m/s.
Find the two possible values for the mass of the second truck.

10 A cannon of mass 400 kg fires a shell of mass 8 kg with an initial velocity of 500 m/s.

Once the cannon has been fired, it recoils through a distance of 1.2 m.
   a) Calculate the initial velocity of the cannon as it recoils.
   b) Calculate the deceleration of the cannon as it recoils.
   c) Find the coefficient of friction between the cannon and the floor.

State any assumptions that have been made in answering this question.
What could be done to the cannon to reduce the recoil distance?

11 A body of mass 100 kg falls freely from rest a distance of 5 metres on to a pile of mass 1000 kg. The body does not rebound and the pile is driven into the ground a distance of 0.2 metres.

   a) Calculate the speed of the body just before the impact with the pile.
   b) Calculate the speed of the combined body and pile just after the impact.
   c) Find the deceleration of the combined body and pile. (Assume the deceleration is constant.)
   d) Find the force resisting the motion of the combined body and pile as it moves through the ground.

12 A, B and C are three particles of masses m, $2m$ and $5m$, respectively. Initially they lie at rest on a smooth horizontal plane with B between A and C and AB = a and BC = $2a$. The particle A then moves towards B with velocity $6u$.

On collision, A and B coalesce to form a single particle, P, which moves on to collide with C. After the collision with C, P has speed $0.5u$ in the same direction as it was originally moving.

Calculate

a) the initial velocity of the combined particle P;

b) the velocity of C after the second collision;

c) the time that elapses from the start of the motion to the second collision.

Having studied this chapter you should know that

● the momentum of a particle of mass m kg moving with velocity v m/s is given by mv

● momentum is a vector quantity and its direction is the same as the direction of the velocity and that the S.I. units for momentum can be expressed as kg m/s or as Ns

● the Conservation of Momentum states that the momentum immediately before a collision is the same as the momentum immediately after a collision provided no external impulses have acted during the collision

● an application of the Conservation of Momentum to a collision problem requires clear diagrams showing the velocities before the collision and the velocities after the collision

REVISION EXERCISE

1 Two particles, A and B, of mass 3 kg and 1 kg, respectively, are moving on a smooth horizontal table in opposite directions with speeds of 6 m/s and 2 m/s, respectively, when they collide directly. Immediately after the collision the speed of B is 1 m/s greater than the speed of A. Find the speed of each particle immediately after the collision.

2 Two particles P and Q of mass $3m$ and $5m$, respectively move towards each other with speeds $6u$ and $2u$. Immediately after the collision the speed of P is $\frac{1}{2}u$. Find the two possible speeds of Q immediately after the collision.

3 A particle P moving with speed u collides directly with an identical particle Q which is at rest. If the speed of Q after the collision is $\frac{2}{3}u$, find the speed of P after the collision.

4 A car of mass 1200 kg tows a second car of mass M kg by means of a tow-rope. Just before the tow-rope tightens, the first car was travelling at 1.5 m/s and the second car was stationary. Just after the tow-rope tightened, both cars have a speed of 0.9 m/s. Find the value of M.

5   A pellet travelling at 100 m/s strikes the centre of a wooden block which is initially at rest on a smooth horizontal surface. The mass of the pellet is 0.12 kg and the mass of the block is 3 kg. The pellet becomes embedded in the block. Find the speed of the block and pellet immediately after the impact.

6   Three smooth spheres, A, B and C, lie at rest on a smooth horizontal table with B between A and C. The masses of A, B and C are $3m$, $4m$ and km, respectively. Sphere A is projected along the surface directly towards B with speed $4u$ and collides with B. A is brought to rest by the collision.
  i)   Find the speed of B after this collision.

After the collision between B and C, B and C are moving in the same direction with speeds $\frac{1}{2}u$ and $3u$.

  ii) Determine the value of k.

7   P and Q are 4 m apart at opposite ends of a smooth plane inclined at 30° to the horizontal. Initially a particle S of mass m is projected up the plane from P with speed 8 m/s and a particle T of mass $2m$ is released from rest at Q.

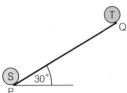

  a) Prove that the particles collide after half a second and find the distance of the particles from P at the instant when they collide.
  b) Find the velocity of each particle just before the collision.

When the particles collide they coalesce.
  c) Find the speed of the combined particle just after the collision and state whether the combined particle is moving up or down the slope.
  d) Find the length of time that elapses from the beginning of the motion until the combined particle reaches P.
  e) Draw a (t, v) graph to illustrate the velocity of S during the motion.

8   Three skaters, P, Q and R, of masses 70 kg, 50 kg and m kg, respectively, are skating along the same straight line and in the same direction. Initially P is skating with speed 4 m/s, Q is skating with speed 1.6 m/s and R is skating with speed 2 m/s.

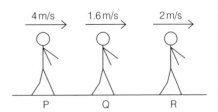

The first collision occurs when P catches up with Q and they then skate as a pair with speed v m/s.
  a) Find the value of v.

After P and Q together bump into R, R's speed increases to 2.75 m/s whilst P and Q continue to skate as a pair and have a speed of 2.5 m/s.
  b) Obtain the value of m.

What assumptions have been made in answering this question?

9   Particles A, B and C, whose masses are m, $2m$ and km, respectively are initially placed at rest on a straight line on a smooth horizontal surface with B between A and C and AB = BC = 3 m.

A is projected towards B with speed 6 m/s.

After the collision between A and B, A is moving in the same direction as before but with speed 1 m/s.

When B and C collide they coalesce and start to move with speed 0.5 m/s.

i)   Find the speed of B immediately after A hits it.

ii)   Find the value of k.

iii)   Draw a (t, x) graph showing the motion of A and the motion of B from the instant when B starts to move and prove that A collides with the coalesced B and C 4.8 seconds after A hit B for the first time.

7 Kinematics 2: Using Calculus to Describe Motion

The purpose of this chapter is to enable you to

- use differentiation and integration to find displacement, velocity and acceleration of particles moving on a straight line

In chapter 4 we saw how motion on a straight line could be illustrated by means of displacement–time and velocity–time graphs and that the gradients and areas under these graphs gave valuable information. In particular:

- the gradient of the displacement–time or (t, x) graph gives velocity;
- the gradient of the velocity–time or (t, v) graph gives acceleration;
- the area under a velocity–time or (t, v) graph gives displacement.

During the course of your Pure Maths studies for C1 and C2 modules, you will have discovered that differentiation provides a quick and neat method of finding the gradient of a curve whose equation is given, whilst integration gives an efficient method of calculating areas under curves. In this chapter the techniques of differentiation and integration are applied to the motion of bodies along straight lines.

Finding Velocity and Acceleration from Displacement

Suppose a particle is moving in a straight line in such a way that its displacement, x metres, from a point O on the line at time t seconds is given by

Time t

$$x = 0.1t^3 + 2t^2 - 8t + 4$$

$$x = 0.1t^3 + 2t^2 - 8t + 4$$

The displacements at various times can be calculated and presented in a table:

t	x
0	4
0.5	0.5125
1	−1.9
1.5	−3.1625
2	−3.2
2.5	−1.9375
3	0.7
3.5	4.7875
4	10.4
4.5	17.6125
5	26.5

and the motion of the particle during the first 5 seconds can be illustrated by a displacement–time graph:

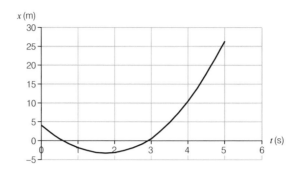

The **velocity**, v m/s, of the particle at an instant can be described as

the gradient of the displacement–time graph at that instant

or

the rate of change of the displacement at that instant

and we know that this can be found by calculating the value of $\dfrac{dx}{dt}$ at that instant.

So, for the example,

$$x = 0.1t^3 + 2t^2 - 8t + 4$$

$$\Rightarrow \quad v = \frac{dx}{dt} = 0.3t^2 + 4t - 8$$

> s is sometimes used to denote displacement. If this is the case,
> $$v = \frac{ds}{dt}$$

This rule gives the velocity for any value of the time t.
The velocity at a particular instant can be determined by substituting the value of t into the formula.

When $t = 3$,

$$v = 0.3 \times 3^2 + 4 \times 3 - 8 = 6.7$$

so the velocity of the particle when $t = 3$ is 6.7 m/s.

The velocity–time graph for this example is shown below:

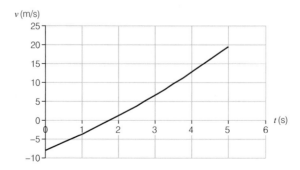

The **acceleration**, a m/s^2, of the particle at an instant can be described as

the gradient of the velocity–time graph at that instant

or

the rate of change of the velocity at that instant

and we know that this can be found by calculating the value of $\dfrac{dv}{dt}$ at that instant.

For the example being considered:

$$x = 0.1t^3 + 2t^2 - 8t + 4$$

$$\Rightarrow \quad v = \frac{dx}{dt} = 0.3t^2 + 4t - 8$$

and a second differentiation gives

$$a = \frac{dv}{dt} = 0.6t + 4$$

The acceleration at a particular instant can be found by substituting the value of t into the formula.

When $t = 3$

$$a = 0.6 \times 3 + 4 = 5.8$$

so the acceleration of the particle when $t = 3$ is 5.8 m/s^2.

The acceleration–time relationship can also be shown graphically in an acceleration–time graph:

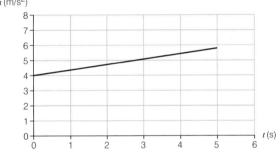

EXAMPLE 1

A particle moves along a straight line in such a way that its displacement, x m, from a fixed point O on the line at time t s is given by

$$x = 0.3t^3 - 3.6t^2 + 10.8t + 2.7$$

a) Find expressions for the velocity and acceleration at time t.
b) Find the times when the particle is at rest and the position of the particle at these times.
c) Describe the motion of the particle during the period $t = 0$ to $t = 10$ and find the average velocity and the average speed of the particle during this interval.

SOLUTION

a) $$x = 0.3t^3 - 3.6t^2 + 10.8t + 2.7$$

$$\Rightarrow \quad v = \frac{dx}{dt} = 0.9t^2 - 7.2t + 10.8$$

$$\Rightarrow \quad a = \frac{dv}{dt} = 1.8t - 7.2$$

EXAMPLE 1 (continued)

b) The particle is at rest when its velocity is zero.

$$v = 0$$
$$\Rightarrow \quad 0.9t^2 - 7.2t + 10.8 = 0$$

You could use your graphical calculator to solve this quadratic equation.

$$[\div 0.9] \quad \Rightarrow \quad t^2 - 8t + 12 = 0$$
$$\Rightarrow \quad (t-2)(t-6) = 0$$
$$\Rightarrow \quad t = 2 \quad \text{or} \quad t = 6$$

When $t = 2$ $x = 0.3t^3 - 3.6t^2 + 10.8t + 2.7 = 0.3 \times 2^3 - 3.6 \times 2^2 + 10.8 \times 2 + 2.7 = 12.3$

and

when $t = 6$ $x = 0.3t^3 - 3.6t^2 + 10.8t + 2.7 = 0.3 \times 6^3 - 3.6 \times 6^2 + 10.8 \times 6 + 2.7 = 2.7$

c) At
$t = 0$,	$x = 2.7$,	$v = 10.8$
$t = 2$,	$x = 12.3$,	$v = 0$
$t = 6$,	$x = 2.7$,	$v = 0$
$t = 10$,	$x = 50.7$,	$v = 28.8$

There cannot be any other changes in direction since the velocity must be zero at the instant when the particle changes its direction of movement. The **only** times when the velocity is zero are $t = 2$ or 6.

so,
starting at $x = 2.7$, from $t = 0$ to $t = 2$, the particle moves in the positive direction and reaches $x = 12.3$ where it reverses direction and moves in the negative direction to reach $x = 2.7$ when $t = 6$. It reverses direction again and moves in the positive direction to reach $x = 50.7$ when $t = 10$.

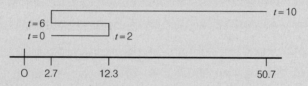

Average velocity during 10 s interval

$$= \frac{\text{change in displacement}}{10} = \frac{50.7 - 2.7}{10} = \frac{48}{10} = 4.8 \text{m/s}$$

Average speed

$$= \frac{\text{distance travelled}}{10} = \frac{(12.3 - 2.7) \times 2 + (50.7 - 2.7)}{10} = \frac{67.2}{10} = 6.72 \text{ m/s}$$

EXERCISE 1

1 A particle is moving along a straight line in such a way that at time t seconds it is x metres away from a fixed point O on the line where

$$x = 2t^3 + 5t^2 + 4$$

a) How far is the particle from O when $t = 2$?

b) What is the velocity of the particle when $t = 2$?

c) What is the acceleration of the particle when $t = 2$?

2 A particle is moving along the line OA so that t seconds after passing O it is x metres from O where

$$x = 6t - \frac{1}{2}t^3$$

a) Find the velocity of the particle when $t = 3$ and when $t = 10$.
b) Find the time at which the particle is momentarily at rest and its distance from O at this time.

3 A car travels in a straight line OA so that its velocity after t seconds is v m/s where

$$v = 4 + 1.5t - \frac{t^2}{80}$$

a) Find the initial velocity of the car.
b) Find the acceleration of the car when $t = 30$.

4 A particle is moving along the line OA in such a way that at time t s it is s m away from O where

$$s = 0.15t^3 - 1.2t^2 + 2.4t$$

a) Find expressions for the velocity and acceleration at time t.
b) Find the times when the particle is at rest and the position of the particle at these times.
c) Describe the motion of the particle during the period $t = 0$ to $t = 5$ and find the average speed and the average velocity of the particle during this interval.

5 A particle moves in a straight line OA so that t seconds after leaving O it is at x metres from O where

$$x = \alpha t^2 + \beta t^3$$

When $t = 12$ the particle is travelling at its **maximum** velocity of 9 m/s.
a) Prove that $24\alpha + 432\beta = 9$.
b) What must the value of $\dfrac{dv}{dt}$ be when $t = 12$? Write down another equation that must be satisfied by α and β.
c) Find α and β and hence find the distance of the particle from O when $t = 12$.

6 A particle of mass 0.8 kg moves on a straight line OU in such a way that at time t s the displacement of the particle from O is x metres where

$$x = 0.3t^2(t^2 - 4)$$

a) Find expressions for the velocity and acceleration of the particle at time t.
b) Find the momentum of the particle when $t = 3$.
c) Find the magnitude of the resultant force acting on the particle when $t = 3$.

7 A particle of mass 3 kg is moving along a straight line OX under the action of a single force, F N, which varies with time.
At time t s the particle's displacement from O is s m where

$$s = 2.6 + 1.5t + 0.9t^2 - 0.1t^3$$

a) Prove that $F = 5.4 - 1.8t$.
b) Find the position and speed of the particle at the instants when the force acting on the particle has magnitude 1.8 N.

Finding Velocity and Displacement from Acceleration

We have seen that if the displacement of a particle moving on a straight line is given as a function of time then differentiating this function gives the velocity and differentiating again gives the acceleration.

Since integration is just the reverse of differentiation, we can work back from the acceleration to the velocity or from the velocity to the displacement using integration.

EXAMPLE 2

A particle moves along a straight line on which there is a fixed point O. Initially the displacement of the particle from O is 2.6 m and it is moving with a velocity of 0.3 m/s. At time t the particle experiences an acceleration of a m/s^2 where

$$a = 0.06t^2 - 0.2$$

Find the velocity and displacement of the particle when $t = 2$.

> The acceleration is **not** constant so we must **not** use the constant acceleration equations!

$$a = \frac{dv}{dt} = 0.06t^2 - 0.2$$

Integrating $\quad \Rightarrow \quad v = \int (0.06t^2 - 0.2) \, dt$

> Don't forget the **constant of integration** that is always present when you do an indefinite integral.

$$\Rightarrow \quad v = 0.02t^3 - 0.2t + c$$

We are told that when $t = 0$, $v = 0.3$ and substituting these values into the last equation will enable the value of the constant to be determined:

$$0.3 = 0.02 \times 0^3 - 0.2 \times 0 + c$$

$$\Rightarrow \quad c = 0.3$$

$$\Rightarrow \quad v = 0.02t^3 - 0.2t + 0.3$$

When $t = 2$,

$$v = 0.02 \times 2^3 - 0.2 \times 2 + 0.3 = 0.06$$

so the velocity of the particle when $t = 2$ is 0.06 m/s.

$$v = \frac{dx}{dt} = 0.02t^3 - 0.2t + 0.3$$

Integrating $\quad \Rightarrow \quad x = \int (0.02t^3 - 0.2t + 0.3) \, dt$

> Again, don't forget the **constant of integration**! Use a different letter (not c) this time to avoid confusion.

$$\Rightarrow \quad x = 0.005t^4 - 0.1t^2 + 0.3t + k$$

We are told that when $t = 0$, $x = 2.6$ so

$$2.6 = 0.005 \times 0^4 - 0.1 \times 0^2 + 0.3 \times 0 + k$$

$$\Rightarrow \quad k = 2.6$$

$$\Rightarrow \quad x = 0.005t^4 - 0.1t^2 + 0.3t + 2.6$$

When $t = 2$,

$$x = 0.005 \times 2^4 - 0.1 \times 2^2 + 0.3 \times 2 + 2.6 = 2.88$$

so the displacement of the particle from O when $t = 2$ is 2.88 m.

The relationship between displacement, velocity and acceleration can be summarised diagrammatically:

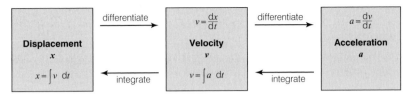

EXAMPLE 3

A body moving along a straight line passes a fixed point O with a velocity of -12 m/s. t seconds later, when it is s metres from O, its acceleration is $6t$ m/s^2. Describe the motion of the body during the next 4 seconds.

To describe the motion of the body, we certainly need to know expressions for the velocity and displacement of the particle in terms of t.

We have

$$a = \frac{dv}{dt} = 6t$$

integrating $\Rightarrow$ $v = \int 6t\, dt$

$\Rightarrow$ $v = 3t^2 + c$

We know that when $t = 0$, $v = -12$

$\Rightarrow$ $-12 = 3 \times 0^2 + c$

$\Rightarrow$ $c = -12$

$\Rightarrow$ $v = 3t^2 - 12$

$$v = \frac{ds}{dt} = 3t^2 - 12$$

integrating $\Rightarrow$ $s = \int (3t^2 - 12)\, dt$

$\Rightarrow$ $s = t^3 - 12\,t + k$

We know that when $t = 0$, $s = 0$

$\Rightarrow$ $0 = 0 + 0 + k$

$\Rightarrow$ $k = 0$

$\Rightarrow$ $s = t^3 - 12\,t$

Using the factorised forms for the velocity and the displacement we can give a full description of the body's motion during the 4 second period.

EXAMPLE 3 (continued)

Since $v = 3t^2 - 12 = 3(t^2 - 4) = 3(t - 2)(t + 2)$
the velocity of the particle is negative for $0 \leqslant t < 2$, zero when $t = 2$ and positive when $2 < t$.
Since $s = t^3 - 12t = t(t^2 - 12)$
the particle passes through O (i.e. $s = 0$) when $t = 0$ or $\sqrt{12}$.

The direction of motion of the particle changes when $v = 0$ and we know this happens when $t = 2$. Substituting $t = 2$ into $s = t^3 - 12t$ gives $s = 2^3 - 12 \times 2 = -16$.

When $t = 4$, $s = 16$ and $v = 36$.

The body's motion can be illustrated in the diagram below:

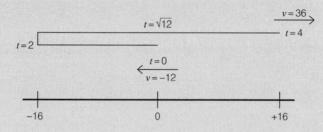

EXAMPLE 4

A particle of mass 2 kg is pulled across a rough horizontal floor by a horizontal string. The coefficient of friction between the particle and the floor is $\dfrac{15}{98}$. Initially the particle is at rest at a point O on the floor and at time t seconds the tension in the string has magnitude $\dfrac{3t^2}{100}$ N.

a) Prove that until $t = 10$ the particle is at rest at O.
b) Find the velocity and displacement of the particle when $t = 15$.

a) The particle will remain at rest at O if the resultant force acting on the particle is zero.

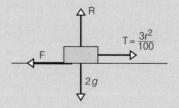

$$[\uparrow] \qquad R - 2g = 0$$
$$\Rightarrow \quad R = 19.6$$

$$[\rightarrow] \qquad T - F = 0$$
$$\Rightarrow \quad F = T$$
$$\Rightarrow \quad F = \frac{3t^2}{100}$$

EXAMPLE 4 (continued)

But F cannot exceed its maximum possible value:

$$\Rightarrow \quad \frac{3t^2}{100} \leqslant \mu N = \frac{15}{98} \times 19.6$$

> Remember that the frictional force has a maximum possible value, F_{max}, given by $F_{max} = \mu R$.

$$\Rightarrow \quad \frac{3t^2}{100} \leqslant \mu N = 3$$

$$\Rightarrow \quad t^2 \leqslant 100$$

$$\Rightarrow \quad t \leqslant 10$$

So, for the first 10 seconds, the forces on the particle are in equilibrium and the particle remains at rest at O. If $t > 10$, then equilibrium is broken since F cannot exceed its maximum value of 3 N.

b) If $t > 10$ then the block accelerates and the force force diagram is shown.

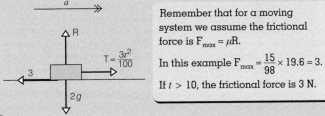

> Remember that for a moving system we assume the frictional force is $F_{max} = \mu R$.
>
> In this example $F_{max} = \frac{15}{98} \times 19.6 = 3$.
>
> If $t > 10$, the frictional force is 3 N.

$$[\text{N2L, } \rightarrow] \quad T - 3 = 2a$$

$$\Rightarrow \quad \frac{3t^2}{100} - 3 = 2a$$

$$\Rightarrow \quad a = \frac{3t^2}{200} - \frac{3}{2} \quad \text{provided } t \geqslant 10$$

Integrating, to find the velocity

$$\Rightarrow \quad v = \int \left(\frac{3t^2}{200} - \frac{3}{2} \right) dt$$

$$\Rightarrow \quad v = \frac{t^3}{200} - \frac{3t}{2} + c$$

We know that when $t = 10$, $v = 0$

$$\Rightarrow \quad 0 = 5 - 15 + c$$

$$\Rightarrow \quad c = 10$$

$$\Rightarrow \quad v = \frac{t^3}{200} - \frac{3t}{2} + 10 \quad \text{provided } t \geqslant 10$$

Integrating, to find the displacement

$$\Rightarrow \quad x = \int \left(\frac{t^3}{200} - \frac{3t}{2} + 10 \right) dt$$

$$\Rightarrow \quad x = \frac{t^4}{800} - \frac{3t^2}{4} + 10t + k$$

We know that when $t = 10$, $x = 0$

EXAMPLE 4 (continued)

$$\Rightarrow \quad 0 = \frac{25}{2} - 75 + 100 + k$$

$$\Rightarrow \quad k = -\frac{75}{2}$$

$$\Rightarrow \quad x = \frac{t^4}{800} - \frac{3t^2}{4} + 10t - \frac{75}{2}$$

When $t = 15$,

$$v = \frac{15^3}{200} - \frac{3 \times 15}{2} + 10 = \frac{135}{8} - \frac{45}{2} + 10 = \frac{35}{8}$$

and

$$x = \frac{15^4}{800} - \frac{3 \times 15^2}{4} + 10 \times 15 - \frac{75}{2} = \frac{225}{32}$$

So, when $t = 15$, the velocity of the particle is $\frac{35}{8}$ m/s and the displacement of the particle from O is $\frac{225}{32}$ m.

EXERCISE 2

1 A particle moves along a straight line in such a way that its velocity, v m/s, at time t seconds after it passes through a fixed point O on the line is given by

$$v = 0.36 + 0.24t - 0.18t^2$$

a) What is the initial velocity of the particle?
b) Find the displacement of the particle from O when $t = 3$.

2 A particle moves along a straight line in such a way that initially it has a displacement of 2 m from a fixed point O on the straight line and is moving with velocity 0.6 m/s. At time t the particle experiences an acceleration of $-0.0018t^2$ m/s^2.
a) Find an expression for the velocity at time t.
b) Find an expression for the displacement of the particle at time t.
c) Prove that the particle is at rest when $t = 10$ and find its displacement at this time.

3 The acceleration, at time t seconds, of a body moving along a straight line is $3t^2 + 5$ m/s^2. The initial velocity of the body is 8 m/s.
Find expressions for the velocity of the body and the displacement of the body from its initial position at time t.
Find the displacement of the body from its initial position when $t = 2$ and when $t = 3$ and hence determine the distance moved by the particle during the third second of the motion.

4 A particle P moves on a straight line so that its velocity v m/s at time t seconds is given by

$$v = 3t^2 - 16t + 5$$

a) Calculate the values of t when P is at rest.
b) Sketch the (t, v) graph for $0 \leqslant t \leqslant 6$.
c) Calculate the acceleration of P when $t = 4$.
d) Given that after t seconds the distance of P from a fixed point O on the line is s metres and that when $t = 0$, $s = 2$, find an expression for s in terms of t.

5 A particle moves on a line in such a way that at time t seconds after passing through a fixed point O on the line with velocity 0.4 m/s the particle has acceleration a m/s^2 given by

$$a = \alpha t^2 + \beta$$

where α and β are constants.
a) Find expressions, involving α, β and t, for the velocity and displacement of the particle at time t.

When $t = 1$ the velocity of the particle is 0.6 m/s and the displacement of the particle from O is 0.2 m.
b) Find the values of the constants α and β.
c) Determine the velocity of the particle when $t = 3$ and the average velocity of the particle between $t = 0$ and $t = 3$.

6 A particle is moving along a straight line in such a way that at time t its velocity v is given by

$$v = 12t - 3t^2.$$

Initially the particle is at a point O on the straight line and at time t the displacement of the particle from O is x.
a) Find an expression for the acceleration, a, at time t.
b) Find an expression for x in terms of t. Hence find the value of t when the particle returns to O.

7 P and Q are two particles which move along parallel grooves lying on the two sides of a straight line.

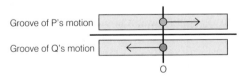

At $t = 0$, P passes a fixed point O on the line with initial velocity 4 m/s and in the subsequent motion P has a constant acceleration of -0.6 m/s^2.

At $t = 0$, Q also passes the point O and has an initial velocity of -2.3 m/s. In the subsequent motion, at time t, Q has an acceleration of $(1.2t + 0.4)$ m/s^2.

Find the time when the two particles are next adjacent to each other and find their velocities at this time.

8 The diagram shows a block of mass 2 kg moving across a smooth table under the influence of two forces: one has constant magnitude 10 N whilst the other has magnitude $4t$ N where t denotes the time that the block has been moving. Initially the block is at rest.

If x denotes the displacement of the block from its starting position at time t and v denotes the velocity at time t;

a) prove that $\dfrac{dv}{dt} = 5 - 2t$;

b) hence find v in terms of t and determine the time when the block is stationary.

c) Find the displacement of the block from its initial position at the time when it is stationary.

9 A and B are particles of mass 2 kg and 3 kg, respectively, which are moving on a straight line passing a point O. Initially, particle A is at rest at O and particle B has a displacement of +2 m from O and is moving with velocity 0.3 m/s.

In the subsequent motion, until the particles collide, A moves along the line under the action of a single force, F_1, which acts along the line in the positive direction and at time t has magnitude $1.2t + 0.8$ N and B moves along the line under the action of a single force, F_2, which acts along the line in the positive direction and at time t has magnitude $1.8t$ N.

a) Find the velocity of A at time t and prove that the displacement of A from O at time t is given by $x_A = 0.1t^3 + 0.2t^2$.

b) Find expressions for the velocity and displacement from O of the particle B at time t.

c) Hence determine when and where the collision between the two particles occurs.

When the particles collide they coalesce to form a single particle.

d) Find the velocity of the combined particle just after the collision.

Using Calculus to Obtain the Constant Acceleration Equations

The equations of motion for a body moving with constant acceleration were introduced in chapter 4 where they were derived from the velocity–time graph for the motion. They can also be readily obtained using the methods of this chapter.

Suppose that a particle is moving on a straight line with constant acceleration a m/s^2 in such a way that initially it is at a point O on the line and moving with velocity u m/s and at time t it is at a displacement x m from O and moving with velocity v m/s.

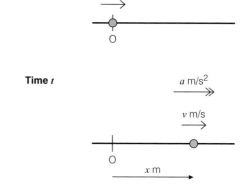

We have

$$\text{acceleration} = \frac{\mathrm{d}v}{\mathrm{d}t} = a$$

integrating $\quad \Rightarrow \quad v = \int a \, \mathrm{d}t$

$\qquad\qquad \Rightarrow \quad v = at + c$

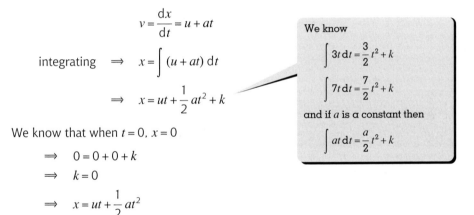

We know

$$\int 3 \, \mathrm{d}t = 3t + c$$

$$\int 7 \mathrm{d}\,t = 7t + c$$

and if a is a constant then

$$\int a \, \mathrm{d}t = at + c$$

To find the value of c we use the fact that when $t = 0$, $v = u$

$\qquad \Rightarrow \quad u = a \times 0 + c$

$\qquad \Rightarrow \quad c = u$

$\qquad \Rightarrow \quad v = at + u$

which we have usually written as

$\quad v = u + at$

We now have

$$v = \frac{\mathrm{d}x}{\mathrm{d}t} = u + at$$

integrating $\quad \Rightarrow \quad x = \int (u + at) \, \mathrm{d}t$

$\qquad\qquad \Rightarrow \quad x = ut + \frac{1}{2} at^2 + k$

We know

$$\int 3t \, \mathrm{d}t = \frac{3}{2} t^2 + k$$

$$\int 7t \, \mathrm{d}t = \frac{7}{2} t^2 + k$$

and if a is a constant then

$$\int at \, \mathrm{d}t = \frac{a}{2} t^2 + k$$

We know that when $t = 0$, $x = 0$

$\qquad \Rightarrow \quad 0 = 0 + 0 + k$

$\qquad \Rightarrow \quad k = 0$

$\qquad \Rightarrow \quad x = ut + \frac{1}{2} at^2$

Integration has given two of the five equations; the remaining three constant acceleration equations may be deduced from these two.

From $\qquad v = u + at$

we obtain $\qquad u = v - at$

so $\qquad\qquad x = ut + \frac{1}{2} at^2$

$\qquad\qquad\qquad = (v - at)t + \frac{1}{2} at^2$

$\qquad\qquad\qquad = vt - at^2 + \frac{1}{2} at^2$

$\qquad\qquad\qquad = vt - \frac{1}{2} at^2$

which gives

$$x = vt - \frac{1}{2} at^2$$

Adding equations

$$x = ut + \frac{1}{2}at^2$$

and

$$x = vt - \frac{1}{2}at^2$$

gives

$$2x = ut + vt \Rightarrow x = \frac{1}{2}(u + v)t$$

so

$$x = \frac{1}{2}(u + v)t$$

Rearranging $v = u + at$ to make t the subject gives

$$t = \frac{v - u}{a}$$

and substituting for t in $x = \frac{1}{2}(u + v)t$ then gives

$$x = \frac{1}{2}(u + v)\frac{(v - u)}{a} = \frac{v^2 - u^2}{2a}$$

$$\Rightarrow \quad 2ax = v^2 - u^2$$

$$\Rightarrow \quad v^2 = u^2 + 2ax$$

We have now derived all five of the constant acceleration equations.

Having studied this chapter you should know how

● to use the methods of differentiation and integration to deduce displacements, velocities and accelerations from knowledge of just one of these quantities using the connections

● to use these methods to obtain the constant acceleration equations

REVISION EXERCISE

1 The velocity v m/s at time t of a particle moving along a straight line is given by the equation

$$v = 9 - t^2 \qquad 0 \leqslant t \leqslant 3$$

a) Show by means of a sketch graph how v varies with t for $0 \leqslant t \leqslant 3$.

b) Find the acceleration of the particle when $t = 2$.

c) Show that the particle travels 18 m during the interval $0 \leqslant t \leqslant 3$.

2 A particle moves on a straight line. The velocity of the particle t seconds after it has passed through a fixed point O on the straight line is v m/s where

$$v = 3.2t - 0.075t^2$$

i) Obtain expressions, in terms of t, for the acceleration, a m/s^2, and the displacement from O, x m, at time t.

ii) Find the displacement of the particle from O at the instant when its acceleration is 2 m/s^2.

3 A particle moves on a straight line in such a way that at time t s its displacement from a fixed point O on the straight line is x m, where

$$x = 0.2t^3 - 2.4t^2 + 7.2t \qquad 0 \leqslant t \leqslant 8$$

a) Obtain expressions involving t for the velocity and acceleration of the particle at time t.

b) Find the times when the particle is stationary.

c) Describe the motion of the particle during each of the periods $0 \leqslant t < 2$, $2 < t < 6$ and $6 < t \leqslant 8$. Sketch the displacement–time graph for the particle.

d) Find the average velocity of the particle during the period $0 \leqslant t \leqslant 8$.

e) Find the average speed of the particle during the period $0 \leqslant t \leqslant 8$.

4 A particle starts from rest at a point O and moves in a straight line until it comes to rest at a point A. The acceleration of the particle t seconds after leaving O is a m/s^2 where

$$a = 0.48 - 0.024t$$

i) Obtain expressions, in terms of t, for the velocity, v m/s, and the displacement from O, x m, at time t.

ii) Prove that OA = 128 m.

iii) Prove that the average speed of the particle is two-thirds of the particle's maximum speed.

5 A particle moves on a straight line. The velocity of the particle t seconds after it has passed through a fixed point O on the straight line is v m/s where

$$v = 0.2t^3 - 0.9t^2 - 6t$$

i) Obtain expressions, in terms of t, for the acceleration, a m/s^2, and the displacement from O, x m, at time t.

ii) Find the displacement from O at the instant when the acceleration is zero.

6 A particle P travels in a straight line from the point O to the point A and back to O. At time t seconds after starting from O, the displacement of P from O is x m, where

$$x = 2t^3 - t^4$$

Find

a) expressions for the velocity and acceleration of P;

b) the value of t at the instant when P returns to O;

c) the speed with which P returns to O;

d) the value of t at the instant when P reaches A;

e) the maximum speed while P is travelling from O towards A.

(OCR Jan 2001 M1)

7 A particle starts from rest at O and travels in a straight line to A. The time taken for the journey from O to A is 25 s, and the particle reaches A with a speed of 15 m/s. The velocity of the particle t seconds after it leaves O is v m/s. It is given that

$$v = kt^2$$

where k is a constant.

a) Show that $k = \dfrac{3}{125}$.

b) Find the distance OA.

c) Find the distance of O from A at the instant when the acceleration is 0.72 m/s².

(OCR Jun 2001 M1)

8 A man P runs in a straight line from O to A, leaving O at time $t = 0$. At time t seconds his velocity is $(2 + 0.006t^2)$ m/s.

a) Find an expression for P's displacement from O in terms of t.

Another man Q leaves O 10 s later than P and runs directly to O at a constant velocity of 5.75 m/s.

b) Find an expression for Q's displacement from O in terms of t, for $t \geqslant 10$.

c) Show that Q is ahead of P when $t = 20$ s.

d) Given that OA = 115 m, find the distance between P and Q at the instant that Q reaches A.

(OCR Jan 2002 M1)

9 A particle P travels in a straight line so that, at time t seconds after leaving a fixed point O, its acceleration is $-\dfrac{1}{10}t$ m/s². At time $t = 0$, the velocity of P is V m/s.

a) Find, by integration, an expression in terms of t and V for the velocity of P.

b) Find the value of V, given that P is instantaneously at rest when $t = 10$.

c) Find the displacement of P from O when $t = 10$.

(OCR May 2002 M1)

10 Particles A and B move along parallel, adjacent straight lines. At time $t = 0$, both particles are adjacent to each other at a point O, with B at rest.
The displacement of A from O at time t seconds is x_A where

$$x_A = 0.01t^3 + 0.025t^2 + 0.04t$$

and the acceleration of B at time t is a_B where

$$a_B = 0.08t + 0.02$$

a) Obtain expressions for the velocity of each particle at time t.

b) Prove that the distance between the two particles at the instant when they have the same velocity is $\dfrac{14}{75}$ m and state which particle is closer to O.

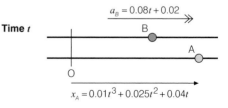

Revise chapter 2 before attempting this exercise.

1   A sledge is pulled by the forces shown in the diagram. Find the magnitude and direction of the resultant force.

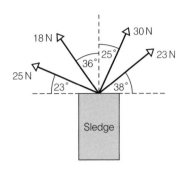

2   Two forces of magnitude 2 N and 3 N act at the point O. The angle between the forces is 40°. The resultant of the two forces has magnitude R N and makes an angle θ with the force of magnitude 2 N, as shown in the diagram. Calculate the values of R and θ.

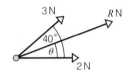

(OCR Jun 2000 M1)

3   A car is being pulled by two boys with ropes as shown in the diagram which also shows the direction of the car's acceleration.

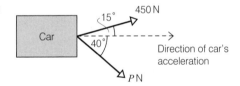

Determine the value of P and the magnitude of the resultant force.

4   ABCD is a square. A force of 5 N acts along AB, a force of 8 N acts along AC and a force of 3 N acts along AD. Find the magnitude and direction of the resultant of these forces.

5   Three coplanar forces of magnitudes 3 N, 8 N and 6 N, with directions as shown in the diagram, act on a particle P. The resultant of these three forces has magnitude R N and acts in the direction of the force of magnitude 8 N. Calculate the value of θ and the value of R.

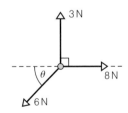

(OCR Nov 1995 M1)

6   Three forces of magnitudes 6 N, 10 N and 18 N act a point P in the directions shown in the diagram.
Find

    a)   the components of the resultant of the three forces

        i)   parallel to the line AB;

        ii)   perpendicular to the line AB;

    b)   the magnitude of the resultant of the three forces;

    c)   the angle that the resultant of the three forces makes with AB.

7   Two forces P and Q have magnitudes 3 N and 6 N, respectively and act at the point O. The angle between P and Q is acute. The resultant of P and Q is R, and R makes an angle of 20° with Q, as shown in the diagram.
Find, in either order,

    a)   the angle between P and Q;

    b)   the magnitude of R.

(OCR Mar 1999 M1)

REVISION 2 Equilibrium

Revise chapter 3 before attempting this exercise.

Note that this exercise will require knowledge of the result

$$W = mg$$

relating the weight, W, of an object to its mass, m, and the acceleration due to gravity, g.

1 Susie uses a strap to pull her suitcase, at a constant speed in a straight line, along the horizontal floor of an airport departure lounge. The strap is inclined at 50° to the horizontal and the frictional force exerted on the case by the floor has magnitude 20 N. Modelling the suitcase as a particle, find the tension in the strap.

(OCR Jun 1996 M1)

2 A book, which may be modelled as a particle of weight 8 N, rests in equilibrium on a desk top inclined at 28° to the horizontal. Find the frictional force acting on the book.

The coefficient of friction between the book and the desk top is 0.6. Determine whether the equilibrium is limiting.

(OCR Jun 1999 M1)

3 A child is sitting on a swing. The child's mother is holding the seat of the swing so that it is in equilibrium. The forces acting on the seat of the swing are as shown in the diagram. The horizontal force has magnitude R, and the force inclined at 40° to the vertical has magnitude T. Find R and T.

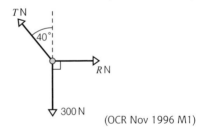

(OCR Nov 1996 M1)

4 A particle is in equilibrium under the action of the three coplanar forces whose magnitudes and directions are show in the diagram. Find the values of P and Q.

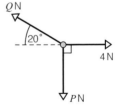

The force of magnitude 4 N is now removed from the system. State the direction in which the particle begins to move.

(OCR Nov 1997 M1)

5 A light inextensible string is attached to the uppermost point of the handle of a loaded basket, which rests on a horizontal shelf. The upper end of the string is attached to a fixed point C, and the string is taut and vertical, as shown in the diagram. The force exerted on the basket by the shelf has magnitude 204 N when the tension in the string is 90 N. Calculate the mass of the loaded basket.

The shelf is lowered so that the basket is no longer in contact with it. Calculate the new tension in the string.

(OCR Mar 1996 M1)

6 A straight path is inclined at an angle of 15° to the horizontal. A loaded skip of total mass 1500 kg is at rest on the path and is attached to a wall at the top of the path by a rope. The rope is taut and parallel to a line of greatest slope of the path, as shown in the diagram. Calculate the normal and frictional components of the contact 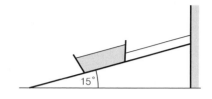 force exerted on the skip by the path when the tension in the rope is 2000 N.

After the rope is cut the skip is on the point of slipping down the path. Calculate the coefficient of friction between the skip and the path.

(OCR Mar 1996 M1)

7 A camping lamp P, of mass 1.2 kg, is supported by two light wires fixed inside a tent. The camping lamp hangs in equilibrium with the wires inclined to the horizontal at angles of 20° and 25°, as shown in the diagram. Find the tension in each wire.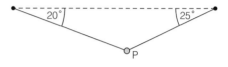

(OCR Nov 1997 M1)

8 A straight footpath makes an angle of α with the horizontal. An object P of weight 1250 N rests on the footpath. The coefficient of friction between the object and the footpath is 0.1. The least magnitude of a force, acting up the footpath, which will hold 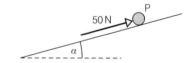 the object at rest on the footpath is 50 N (see diagram). By treating the object as a particle, show that the value of α satisfies

$$10 \sin \alpha - \cos \alpha = 0.4$$

(OCR Mar 1997 M1)

9 A small smooth ring R, of mass 0.3 kg, is threaded on a light inextensible string whose ends are attached to two fixed points A and B which are at the same horizontal level. A force of magnitude X N is applied to the ring in a direction parallel to AB, as shown in the diagram. When the ring is in equilibrium with both parts of the string taut, angle BAR = 30° and angle ABR = 40°. Find the tension in the string and the value of X.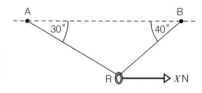

(OCR Jun 1996 M1)

10 A small toy has mass 1.2 kg. When a child tries to move the toy along a horizontal floor by pushing with a force of magnitude 5 N, acting downwards at an angle of 30° to the horizontal as shown in Fig. 1, she is unsuccessful. When the child tries to move the toy by pulling with a force of magnitude 5 N, acting upwards at an angle of 30° to the horizontal as shown in Fig. 2, she is successful. Show that the coefficient of friction between the toy and the floor lies between 0.30 and 0.47, correct to 2 significant figures.

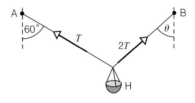

Fig. 1   **Fig. 2**

(OCR Mar 1998 M1)

11 A hanging flower basket H, of weight 50 N, is held in equilibrium by two light inextensible strings. One string is attached to a fixed point A and this string makes an angle of 60° with the vertical; the other string is attached to a fixed point B and this string makes an angle of θ with the vertical, as shown in the diagram. Given that the tension in the string attached to A is T and the tension in the string attached to B is $2T$, find the values of θ and T.

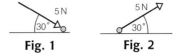

(OCR Jun 1999 M1)

12 On a ski-slope a skier of mass M kg uses a ski-lift which tows him up the slope. The lift stops on a section of the slope that is inclined at 15° to the horizontal. The skier is held in equilibrium, with the force exerted on the skier by the towing bar making an angle of 30° with the slope (see diagram). The skier is modelled as a particle. Given that the contact between the skier and the slope is smooth, and that the force exerted on the skier by the slope has magnitude 600 N, find

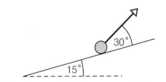

i) the magnitude of the force exerted on the skier by the towing bar,

ii) the value of M.

(OCR Jun 2000 M1)

13 A smooth ring R, of mass 0.6 kg, is threaded on a light inextensible string. One end of the string is attached to a fixed point A and the other end is attached to a ring B, of mass 0.2 kg, which is threaded on a fixed rough horizontal wire which passes through A (see diagram). The system is in equilibrium, with B on the point of slipping and with the part AR of the string making an angle of 60° with the wire.

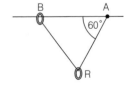

i) Explain, with reference to the fact that ring R is smooth, why the part BR of the string is inclined at 60° to the wire.

ii) Show that the normal contact force between B and the wire has magnitude 4.9 N, correct to 2 significant figures.

iii) Find the coefficient of friction between B and the wire.

(OCR Jun 1998 M1)

14 A parcel, of mass 5 kg, rests on a rough plane inclined at 45° to the horizontal. The parcel is held in equilibrium by a force of magnitude 20 N inclined at 30° to a line of greatest slope of the plane, as shown in the diagram. Draw a diagram showing all the forces acting on the parcel, and show that the normal component of the contact force between the parcel and the plane has a magnitude of 24.6 N, correct to 3 significant figures.

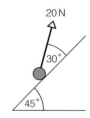

Given that the equilibrium is limiting, with the parcel on the point of moving down the plane, find the coefficient of friction between the parcel and the plane.

(OCR Nov 1996 M1)

15 Particles P and Q, of masses 0.5 kg and 0.6 kg, respectively, are connected by a light inextensible string. Particle P rests on a fixed rough plane inclined at 40° to the horizontal, particle Q hangs freely, and the string passes over a small smooth pulley at the top of the plane, as shown in Fig. 1. Given that the system is in limiting equilibrium, show that the coefficient of friction between P and the plane is 0.73, correct to 2 significant figures.

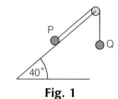

Fig. 1

Particle Q is now replaced by another particle, of mass m kg. A horizontal force of magnitude 4 N is applied to P. This force is in the vertical plane containing the string and is directed away from the inclined plane, as shown in Fig. 2. The system is in limiting equilibrium with P about to move down the inclined plane. Find the value of m.

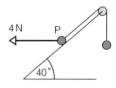

Fig. 2

(OCR Jun 1999 M1)

Revise chapters 4 and 7 before attempting this exercise.

1 A car moving in a straight line with constant acceleration 0.6 m/s² increases its speed from u m/s to $1.5u$ m/s over a distance of 8 m. Calculate the value of u.

(OCR Nov 1995 M1)

2 A coin is thrown vertically upwards, with speed 5 m/s, from the top of a wishing-well. There is no water in the well and the coin hits the bottom of the well 3 s after being thrown. Modelling the coin as a particle, and ignoring air resistance, calculate the depth of the well to the nearest metre.

(OCR Mar 1999 M1)

3 The steam catapult on an aircraft carrier gives an aircraft a constant acceleration of 20 m/s². The speed increases from 10 m/s to 50 m/s in time T seconds, and during this time the aircraft moves a distance of d metres. Find T and d.

(OCR Jun 1995 M1)

4 A particle of mass 0.4 kg moves on a straight line in such a way that at time t its displacement from a fixed point O on the straight line is x metres where

$$x = 0.03t^3 - 0.3t^2 + 0.75t$$

Find

a) the momentum of the particle when $t = 3$;

b) the magnitude of the resultant force acting on the particle when $t = 3$.

5 A ball is projected vertically upwards, from a point O, with speed 32 m/s. Ignoring air resistance, find the time it takes the ball to return to O.

Sketch the (t, v) graph for this motion.

(OCR Jun 1996 M1)

6 A motorist travelling at u m/s joins a straight motorway. On the motorway she travels with a constant acceleration of 0.07 m/s² until her speed has increased by 2.8 m/s.

i) Calculate the time taken for this increase in speed.

ii) Given that the distance travelled while this increase takes place is 1050 m, find u.

(OCR Mar 1998 M1)

7 A snooker ball is moving in a straight line. Its displacement is x metres at time t seconds. For $0 \leqslant t \leqslant 5$ the (t, x) graph consists of two straight line segments as shown in the diagram. Find the velocity of the snooker ball

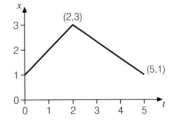

i) when $t = 1$,

ii) when $t = 3$.

Describe briefly what could have happened when $t = 2$.

(OCR Mar 1995 M1)

8 A particle moving along a straight line passes through a fixed point O on the straight line at $t = 0$ and, at this instant, the particle has velocity 9 m/s. At time t the acceleration of the particle is $-0.06t$ m/s^2.

a) Find an expression for the velocity of the particle at time t.

b) Find an expression for the displacement of the particle from O at time t.

c) Verify that the particle returns to O when $t = 30$ and find the speed of the particle at this instant.

d) Find the maximum displacement of the particle from O during the period $0 < t < 30$.

9 A ball is projected vertically upwards from the point A with speed u m/s. The ball returns to A after 8 s. Ignoring air resistance, find the value of u.

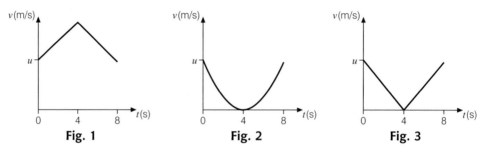

Fig. 1   **Fig. 2**   **Fig. 3**

Students were asked to draw a graph of velocity against time to model the motion of the ball. Three of the (t, v) graphs presented are shown above.

i) By considering the shape of the curve in Fig. 2, explain why this graph is incorrect.

ii) If you think either Fig. 1 or Fig. 3 is correct, state which one. If you think neither is correct, sketch the correct graph.

(OCR Jun 1998 M1)

10 The diagram shows the (t, v) graph for a miniature train as it moves along a straight track. At time $t = 0$ the train passes a point A and is moving at 3 m/s. The furthest point from A reached by the train in the 2 minute period is P. Find

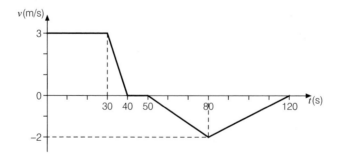

i) the value of t at the instant the train reaches P,

ii) the magnitude of the acceleration of the train in the time interval $50 < t < 80$,

iii) the distance of the train from A at the end of the 2 minute period.

(OCR Mar 1999 M1)

11 A light aircraft starts from rest and travels along a straight level runway until it has sufficient speed for take-off. During its motion along the runway the aircraft travels s metres in t seconds, where

$$s = 0.4t^3 - 0.02t^4$$

Given that at the instant of take-off the horizontal acceleration of the aircraft is zero, calculate
a)   the time taken from rest until take-off;
b)   the distance that the plane travels along the runway.

12 A particle A is projected vertically upwards from a point O on horizontal ground with speed 20 m/s. At the same instant, a particle B is released from rest at a point P which is 30 m above the ground. The point P is not directly above O. Ignoring air resistance, find
i)   the time that elapses between B's arrival on the ground and A's arrival on the ground,
ii)   the speeds of A and B at the instant when they are at the same height, and show that A and B are moving in opposite directions at this instant.

(OCR Mar 1996 M1)

13 A and B are two points 100 metres apart on a road.
At the same instant Peter cycles through A, travelling towards B with speed 8 m/s and constant acceleration of 0.5 m/s^2, and Rosie cycles through B travelling towards A with speed 4 m/s and constant acceleration of 0.3 m/s^2.

Find the time that elapses from this instant to the instant when Peter and Rosie meet and find the distance of their meeting point from A.

14 A particle of mass 2 kg moves along a straight line AB under the action of a single force P which at time t has magnitude $0.3\sqrt{t}$ and acts in the direction $\overrightarrow{\mathbf{AB}}$. At $t = 0$ the particle passes through A and is moving towards B with speed 5 m/s. At $t = 8$ the particle reaches B. Calculate the distance AB.

15 A car can accelerate from 0 to 90 km/hr in 8 seconds.
A car will travel approximately 50 metres in braking from 90 km/hr to a standstill.
a)   Estimate the acceleration, in m/s^2, of a car.
b)   Estimate the retardation of a car when it is braking.

Speed bumps are to be placed on a road.

The maximum speed that a car can safely cross the speed bumps at is 9 km/hr.
The local council wishes to place the speed bumps close enough to restrict a car to a maximum speed of 36 km/hr as it travels between the speed bumps. They believe that the speed–time graph for a reckless driver travelling between the bumps will have the shape shown in the graph.

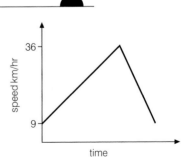

c)   Using your answers to (a) and (b) determine the distance that there should be between bumps to ensure that the reckless driver will not exceed 36 km/hr.
d)   Comment critically on the assumptions made in modelling this situation and suggest briefly how a better model might be made.

16 The line AB is 100 metres long.
At $t = 0$, a particle P leaves A and moves towards B with initial speed 5 m/s and acceleration 2 m/s² and at the same instant a particle Q leaves B and travels towards A with initial speed 8 m/s and acceleration 1 m/s².
Find where and when P and Q meet.

17 A particle moves in a straight line with initial velocity u m/s and constant acceleration a m/s².

    a) Write down expressions involving u and a for

        i) the distance travelled by the particle in the first 3 seconds of the motion;

        ii) the distance travelled by the particle in the first 4 seconds of the motion.

    b) The particle travels 21 metres during the fourth second of the motion. Prove that

$$u + \frac{7a}{2} = 21$$

    c) If the particle travels 37 metres in the eighth second of its motion find another equation connecting u and a.

    d) Find the values of u and a.

18 After stopping at a station, a train accelerates from rest at a m/s² and passes three sets of signals, P, Q and R, where PQ = 240 m and QR = 150 m.
The train passes P with speed u m/s; 20 s later it passes Q and after another 10 s it passes R.
Find

    a) the values of u and a;

    b) the distance of the signal P from the station.

19 Fig. 1 shows a snooker table with the points A, B and C in a straight line, parallel to two sides of the table. A and C are points at opposite ends of the table. A snooker ball is projected from B towards C with speed 4.45 m/s, and after striking the cushion at C it travels to A where it comes to rest. The (t, v) graph for this motion is shown in Fig. 2 below. The graph consists of two straight line segments.

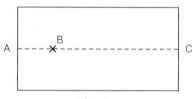

Fig. 1

Calculate the deceleration of the ball during the first stage, B to C, of the motion.

Assuming the diameter of the ball is small, calculate approximations for the length of the table and the distance AB.

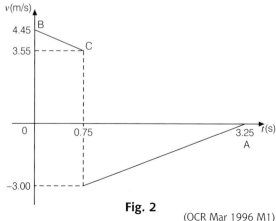

Fig. 2

(OCR Mar 1996 M1)

20 A car starts from rest and its motion, as the driver uses the gears, may be modelled in the following way. The car moves with constant acceleration $2a$ m/s^2 as its speed increases from 0 m/s to 4 m/s. It moves with constant acceleration $1.25a$ m/s^2 as its speed increases from 4 m/s to 9 m/s, and with constant acceleration a m/s^2 as its speed increases from 9 m/s to 23 m/s. Find, in terms of a, the total time taken. Given that the total time taken is 25 s, find the value of a.

Find the total distance moved as the car accelerates from rest to 23 m/s.

(OCR Mar 1995 M1)

21 A car travelling on a straight road accelerates uniformly at a m/s^2 from its initial speed of 20 m/s until it reaches a speed of 30 m/s. The car travels at this speed for 600 m and then decelerates uniformly at $\frac{1}{2}a$ m/s^2 until it comes to rest. The (t, v) graph for the motion is shown. Show that the time spent decelerating is six times as long as the time spent accelerating.

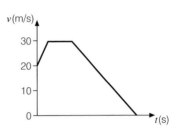

The total distance travelled by the car from time $t = 0$ until it comes to rest is 2900 m. Calculate

i) the time for which the car is accelerating,

ii) the value of a,

iii) the speed of the car 40 s before it comes to rest.

(OCR Nov 1995 M1)

22 A cyclist starts from rest at a set of traffic lights and rides along a straight road until she comes to a stop at a second set of traffic lights. Her velocity, v m/s, after leaving the first set of traffic lights is given by

$$v = \frac{t(60 - t)}{75}$$

a) Find the acceleration of the cyclist when $t = 20$.

b) How long does it take the cyclist to reach the second set of traffic lights?

c) What is the distance between the two sets of traffic lights?

Revise chapter 5 before attempting this exercise.

1 A car of mass 850 kg is moving, with an acceleration of 1.4 m/s², along a straight horizontal road. The engine of the car produces a total forward force of magnitude X N and there is a horizontal resisting force of magnitude 450 N. Find X.

(OCR Jun 1997 M1)

2 A load P, of mass 200 kg, is suspended by the vertical cable of a crane. Another load Q, of mass 120 kg, is suspended from P by another vertical cable. Both cables may be considered as light and inextensible, and any resistances to motion may be neglected. Find the tension in the vertical cable supporting the load P,

i) when the loads are hanging in equilibrium,

ii) when the crane is raising the loads with a vertically upward acceleration of 0.4 m/s².

(OCR Nov 1997 M1)

3 Two bodies, of masses 3 kg and 5 kg, are attached to the ends of a light inextensible string. The string passes over a smooth fixed pulley and the particles are moving vertically with both vertical parts of the string taut. Find the tension in the string.

(OCR Mar 1995 M1)

4 A coin of mass 8 grams is placed flat on a rough board, which is inclined at an angle 25° to the horizontal. The coin moves downwards with acceleration 1.5 m/s². Find the coefficient of friction between the coin and the board.

(OCR Mar 1995 M1)

5 The diagram shows a small parcel of mass 2 kg resting on the seat of a car.

a) Suppose the car is at rest. Draw a diagram to show the forces acting on the parcel.

b) Suppose the car is accelerating forwards at 3 m/s² and that the parcel is at rest relative to the seat.

i) Draw a diagram to show the acceleration of the parcel and the forces acting on the parcel.

ii) Prove that the coefficient of friction, μ, between the parcel and the seat must satisfy the inequality

$$\mu \geqslant \frac{15}{49}$$

6 Two particles P and Q, of masses 2 kg and m kg, respectively, are connected by a light inextensible string. Particle P is held on a smooth horizontal table. The string passes over a smooth pulley R fixed at the edge of the table, and Q is at rest vertically below R (see diagram). When P is released the acceleration of each particle has a magnitude 0.8 m/s². Assuming that air resistance may be ignored, find the tension in the string and the value of m.

(OCR Nov 1996 M1)

7 A schoolboy slides a box, of mass 6 kg, down a straight path inclined at 20° to the horizontal. The initial speed of the box is 5 m/s, and the coefficient of friction between the box and the path is 0.8. Assuming constant acceleration, find the distance travelled before the box comes to rest.

(OCR Jun 1998 M1)

8 A wooden box is pulled along a rough horizontal floor by means of a constant force of magnitude 150 N acting at an angle of 40° above the horizontal. The box may be modelled as a particle of mass 45 kg, and air resistance may be neglected. Draw a diagram showing all the forces acting on the box, and show that the normal component of the contact force of the floor on the box is approximately 345 N.

The coefficient of friction between the box and the floor is 0.3. Calculate the time taken for the box to move 50 m from rest.

(OCR Jun 1997 M1)

9 A car of mass 1000 kg is connected to a trailer of mass 600 kg by a horizontal light rigid tow-bar. The car is pulling the trailer along a straight horizontal road at 10 m/s when the engine is turned off. The car comes to rest, without using the brakes, in a distance of 160 m. During the motion there is a horizontal resisting force of constant magnitude 300 N acting on the car, and a horizontal resisting force of constant magnitude R N acting on the trailer.

i) By considering the motion of the whole system, or otherwise, find the value of R.

ii) Find the tension in the tow-bar.

(OCR Jun 2000 M1)

10 A sledge of mass M kg is pulled up a straight icy path inclined at 17° to the horizontal. The force pulling the sledge is constant, having magnitude 30 N and acting at 57° to the horizontal (i.e. at 40° to the path). The magnitude of the normal contact force between the sledge and the path is 45 N. Ignoring air resistance, and modelling the surface of the icy path as smooth, find the value of M, correct to 3 significant figures.

The sledge is pulled from rest at the bottom of the path and reaches the top of the path in 6 s. Find the length of the path.

(OCR Mar 1999 M1)

11 A crate P containing blocks of wood has a total mass of M kg ($M > 85$). The crate is connected to a counterweight Q, of mass 85 kg, by a rope which passes over a fixed pulley. Initially the counterweight is held in contact with the ground. The crate rests in equilibrium at a height of 18 m above the ground, as shown in the diagram. The counterweight is released. In the ensuing motion both the crate and the counterweight move vertically. The following modelling assumptions are made:

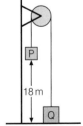

the crate and the counterweight are considered as particles;
the rope is light and inextensible;
the pulley is smooth;
there is no air resistance.

Given that the crate reaches the ground in 7.5 s,

i) show that the acceleration of the crate is 0.64 m/s,

ii) calculate the value of M.

(OCR Jun 1996 M1)

12 Two trucks A and B, of masses 1000 kg and 800 kg, respectively, are connected by a horizontal coupling. An engine pulls the trucks along a straight horizontal track by exerting a horizontal force of magnitude X N on truck A (see diagram). The resistances to the motion of truck A, excluding the tension in the horizontal coupling, may be modelled by a constant horizontal force of magnitude 300 N; for truck B the resistances may be modelled by a constant horizontal force of magnitude 100 N.

a) Given that the trucks are moving with constant speed, find

    i) the tension in the horizontal coupling between the trucks,

    ii) the value of X.

b) Given instead that $X = 800$, find the common acceleration of the trucks.

(OCR Nov 1996 M1)

13 Two loads P and Q, of masses 2 kg and 3 kg, respectively, are connected by a light inextensible string. Load P is on a rough horizontal table. The string passes over a smooth pulley fixed at the edge of the table, and load Q hangs vertically. A force of magnitude 30 N acting at 30° to the table is just sufficient to keep P in equilibrium (see diagram). This force acts in the vertical plane containing the string. Modelling the loads as particles, show that the coefficient of friction between P and the table is approximately 0.743.

The force of magnitude 30 N is now removed. Ignoring air resistance, find the tension in the string in the ensuing motion.

(OCR Jun 2000 M1)

14 A girl sitting on a wooden board slides down a line of greatest slope, which is inclined at 10° to the horizontal, on a snow-covered mountain. The combined mass of the girl and the board is 65 kg, and the magnitude of the frictional force between the board and the slope is 125 N. Air resistance may be ignored. Show that the coefficient of friction between the board and the slope is 0.20, correct to 2 significant figures, and verify that the girl and the board are slowing down.

The girl passes a point A travelling at 5 m/s. Calculate her speed at the point B, where B is 40 m down the slope from A.

Later in the day the girl, still sitting on the board, is pulled up the same slope, with constant speed, by a rope inclined at 30° above the horizontal. The surface of the slope may now be assumed to be smooth. Calculate the magnitude of the force exerted on the board by the slope.

(OCR Jun 1996 M1)

15 A lift has a maximum safe speed of 10 m/s. The (t, v) graph above, which consists of three straight line segments, models the motion of the lift as it moves vertically upwards between two floors, which are 60 m apart, in 12 s. On this journey the highest speed reached is V m/s. Show that the lift does not exceed its maximum safe speed on this journey.

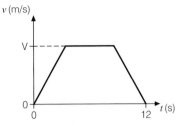

It is given that $V = 8$ and that the time spent accelerating is equal to the time spent decelerating.

i) Find the magnitude of the acceleration.

ii) Show on a sketch the shape of the (t, x) graph for this motion, from the time the lift starts until it is about to start slowing down. [You are not required to carry out any further calculations.]

iii) A trunk is resting on the floor of the lift. When the lift is moving at a constant speed the vertical force exerted on the trunk by the floor of the lift has magnitude 480 N. Find the magnitude of the vertical force exerted on the trunk during the first part of the journey while the lift has an upwards acceleration.

(OCR Nov 1996 M1)

16 Two particles P and Q, of masses 4 kg and 6 kg, respectively, are connected by a light inextensible string which passes over a light smooth pulley A. Particle P is held at rest on a rough horizontal table and particle Q rests on a smooth plane inclined at 30° to the horizontal, as shown in the diagram. The string is taut and lies in a vertical plane perpendicular to the line of intersection of the table and the inclined plane. The particles are released from rest and in the subsequent motion each particle has an acceleration of magnitude 1.9 m/s², provided P has not reached A. Find the tension in the string and the coefficient of friction between P and the table.

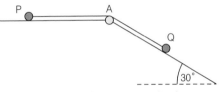

One second after the system is released from rest the string breaks, and P subsequently comes to rest before reaching A. At the instant when the string breaks P is 0.8 m away from A. Find the distance from A at which P comes to rest.

(OCR Mar 1999 M1)

17 The diagram shows three blocks, A, B and C, of masses $8m$, $3m$ and $5m$, respectively.

The block C is initially at rest on a smooth table and is connected to A and B by means of strings passing over small pulleys as shown.

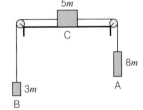

a) Draw separate diagrams showing
　　i) the acceleration of A and the forces acting on A;
　　ii) the acceleration of B and the forces acting on B;
　　iii) the acceleration of C and the forces acting on C.

b) Write down equations relating the acceleration of each block to the forces acting on each block.

c) Hence prove that the acceleration of the system is $\dfrac{5}{16}g$ and find the tension in each string.

d) What assumptions have you made in answering this question?

18 Two trucks A and B, of masses 6000 kg and 4000 kg, respectively, are connected by a horizontal coupling. An engine pulls the trucks along a straight horizontal track, exerting a constant horizontal force of magnitude X N on truck A (see diagram). The resistance to motion for truck A may be modelled by a constant horizontal force of magnitude 360 N; for truck B the resistance may be modelled by a constant horizontal force of magnitude 240 N. Given that the tension in the coupling is T N and that the

acceleration of the trucks is a m/s², show that $T = \frac{2}{5}X$, and express a in terms of X.

Given that the trucks are slowing down, obtain an inequality satisfied by X.

The model is changed so that the resistance for truck B is modelled by a constant force of magnitude 200 N. The resistance for truck A remains unchanged. For this changed model find the range of possible values of X for which the force of the coupling is compressive (i.e. the force in the coupling acting on B is directed from A to B).

(OCR Mar 1997 M1)

19 The diagram shows two particles P and Q, of mass 4 kg and 1 kg, respectively, connected to each other by a string passing over a smooth pulley at C. The particle P rests on a horizontal surface ABC while Q hangs vertically below C.

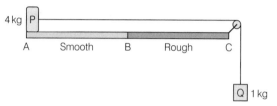

AB is 3 metres long and the surface is smooth.
BC is 2 metres long and the coefficient of friction between P and this surface is μ.
The system is released from rest with the particle P at A and the string taut.
a) For the part of the motion whilst P is moving over AB
   i) draw a diagram to show the forces acting on Q and the acceleration of Q;
   ii) draw a diagram to show the forces acting on P and the acceleration of P;
   iii) find the acceleration of P;
   iv) find the speed of P as it reaches B.
b) If $\mu = 0.8$, for the part of the motion when P is moving over BC
   i) find the acceleration of the system;
   ii) prove that P comes to rest before it reaches C.
c) What would be the value of the coefficient of friction between P and the surface BC if P moves with constant speed over BC?

20 Particles A and B, of masses 0.2 kg and 0.1 kg, respectively, are joined by a light inextensible string. Particle A is placed on a fixed smooth plane inclined at 10° to the horizontal, and is held at rest by a force of magnitude X N which acts in a direction parallel to a line of greatest slope of the plane. The string passes over a smooth pulley P fixed at the bottom of the plane, and the part PB of the string hangs vertically, as shown in the diagram. Find X.

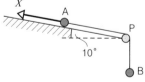

The force of magnitude X N is now removed. Ignoring air resistance, find the tension in the string in the subsequent motion.

The resultant force exerted on the pulley by the string acts in a direction $\theta°$ below the horizontal. State the value of θ.

(OCR Jun 1998 M1)

Revise chapter 6 before attempting this exercise.

1   Two marbles A and B, of mass 4 grams and 5 grams, respectively have the same radius. The marbles move directly towards each other on a smooth horizontal table. Just before they collide A has a speed of 0.6 m/s and B has a speed of 0.4 m/s. After the collision the marbles move in opposite directions, each with speed v m/s. Calculate the value of v.

(OCR Nov 1999 M2)

2   The diagram shows particles A, B and C, of masses $2m$, m and $2m$ respectively, which are free to move in a smooth straight horizontal groove, with B between A and C. Particles A and B are each moving towards C with speeds $3u$ and $2u$, respectively when A strikes B.

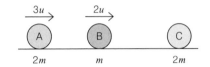

a)   Given that B's speed immediately after the collision is $3u$, find A's speed immediately after the collision.

B subsequently strikes C which is stationary and C starts to move with speed $1.5u$.

b)   Prove that B is brought to rest after this collision.

The time interval between the first and second collisions is T. A third collision takes place.

c)   Find, in terms of T, the time interval between the second and third collisions.

(OCR Jun 2000 M2, adapted)

3   Two spheres, A and B, of masses 0.15 kg and m kg, are moving horizontally on a smooth plane. They move towards each other, in the same straight line, with speeds of 5 m/s and 4 m/s, respectively. On impact they coalesce and the combined particle moves at a speed of 1.4 m/s. Find the two possible values of m.

4   A particle of mass 1.2 kg moving at 5 m/s collides head on with a stationary particle of mass 1.8 kg. Immediately after the collision the particles are moving in the same direction and the speed of the heavier particle is 0.5 m/s greater than the speed of the lighter particle. Find the speed of each particle after the collision.

5   Particles A and B, of masses 3 kg and m kg, respectively, are moving horizontally on a smooth plane. They move towards each other, in the same straight line, with speeds of 1.2 m/s and 0.9 m/s, respectively. When they collide, each particle is brought to rest. Determine the value of m.

6   Particles A and B, of masses 0.3 kg and 0.6 kg, respectively can slide along a straight horizontal groove, PQR. Initially A is at rest at P and B is at rest at Q. The portion of the groove between P and Q is smooth whilst the portion of the groove between Q and R is rough and the coefficient of friction between B and this portion of the groove is $\dfrac{1}{49}$.

At $t = 0$, A is projected towards Q with speed 1.2 m/s. After 2 seconds of motion A hits B and is brought to rest by the collision.

a)   Find the speed of B immediately after the collision.

b)   Prove that as B moves along the groove it experiences a deceleration of 0.2 m/s².

c)   Given that B just reaches the end R of the groove, determine the length of the groove.

7 A particle P of mass $2m$ moves on a straight line with a constant acceleration of 0.75 m/s². At $t = 0$ P passes through a point O on the line with velocity 6 m/s. When $t = 2$ a second particle, Q, of mass 3 kg starts moving along the line from O: when $t = 2$ this particle is at rest and in the subsequent motion has acceleration 4 m/s².

a) Find the time when the two particles collide.

b) Find the velocity of each particle just before the collision.

When the particles collide they coalesce.

c) Find the velocity of the combined mass just after the collision.

8 A rocket of mass $10m$ is flying in a straight line with speed u when it explodes into two fragments. Immediately after the explosion, one fragment is moving with speed $2u$ in the same direction as the original motion of the rocket whilst the other portion is moving with speed $3u$ in the opposite direction to the original motion of the rocket. Find the mass of each of the two fragments.

Preliminary note

Students should be aware that:

- Calculations should yield exact answers or answers that are correct to 3 significant figures, as appropriate, unless otherwise indicated in the question.
- Calculations, in which intermediate values are rounded to 3 significant figures for subsequent use, will not necessarily give sufficiently accurate answers.
- The value of g to be used is 9.8, where g ms^{-2} is the acceleration due to gravity.

1 A particle slides down a line of greatest slope of a smooth plane inclined at an angle of $\alpha°$ to the horizontal. The particle passes through the points A and B with speeds 1 ms^{-1} and 1.5 ms^{-1} respectively. Given that the distance AB is 1.25 m, find

   i) the acceleration of the particle, [3]

   ii) the value of α. [2]

2 A small object falls vertically from rest at a point 0. There is no resistance to the object's motion during the first 1.5 s. At time $t = 1.5$, where t is the time in seconds after the object starts at 0, the object hits the surface of liquid in a reservoir. The object continues to move vertically downwards. While in the liquid and until the object hits the bottom of the reservoir at time $t = 2$, the object's deceleration is constant and equal to 20 ms^{-2}.

   i) Sketch the (t, v) graph for the object's motion, for $0 < t < 2$, where v ms^{-1} is the downward velocity of the object at time t s. [2]

   ii) Find the velocity with which the object hits the bottom of the reservoir. [3]

   iii) Find the depth of liquid in the reservoir. [3]

3 Three coplanar forces, of magnitudes 4N, 5N and PN, act at a point. The directions of the forces are as shown in the diagram.

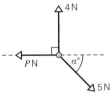

   i) Given that the forces are in equilibrium, find the value of P. [4]

   ii) Given instead that $P = 2.8$ and $\alpha = 30$, find the magnitude of the resultant of the three forces. [5]

4 A block of mass 0.4 kg is on a horizontal table. The block is acted upon by a force of magnitude 7N, which makes an angle of 14° with the horizontal.

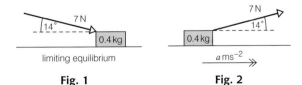

Fig. 1   Fig. 2

   i) When the force acts downwards (see Fig. 1) the block is in limiting equilibrium. Find the coefficient of friction between the block and the table. [5]

   ii) When the force acts upwards (see Fig. 2) the block moves with acceleration a ms^{-2}. Find the value of a. [5]

5 A block B and a particle P are attached to the ends of a light inextensible string. The mass of P is 0.2 kg. The string passes over a small smooth pulley which is fixed at the edge of a rough horizontal table. The string is taut with B held on the table and with P hanging vertically below the pulley (see diagram). B is now released and P moves downwards until it reaches the floor. While P is moving downwards the frictional force acting on B is F N and the tension in the string is T N.

i) By applying Newton's second law to P show that $T < 1.96$. [3]

ii) State, giving a reason, whether F is greater than T, equal to T or less than T. [2]

The mass of B is 0.5 kg and the acceleration of P is 0.8 ms^{-2}.

iii) Find the values of F and T. [3]

Initially P is closer to the floor than B is to the pulley.

iv) Find the deceleration of B during the period after P reaches the floor and before B reaches the pulley. [2]

6 A particle A starts from rest at 0 and travels in a straight line, coming to rest again after 8 s. The velocity of the particle at time t s after starting is v ms^{-1}, for $0 \leqslant t \leqslant 8$, where $v = 2.4t - 0.3t^2$. A particle B starts from rest at 0 at the same instant as A and travels in the same direction along the same straight line. Particle B has constant acceleration a ms^{-2} for $0 < t < 4$ and constant acceleration $-a$ ms^{-2} for $4 < t < 8$.

i) Find the greatest velocity of A during its motion. [4]

ii) Given that the displacements of A and B are the same when $t = 8$, find the greatest velocity of B during its motion. [6]

iii) Find the value of t, in the interval $0 < t < 4$, for which the acceleration of A is the same as the acceleration of B. [3]

iv) Hence or otherwise write down the value of t, in the interval $4 < t < 8$, for which the acceleration of A is the same as the acceleration of B. [1]

7 Three small spheres A, B and C are free to move along a fixed straight line on a horizontal table, with B between A and C. When in motion each of the spheres has deceleration 0.5 ms^{-2}. The masses of A, B and C are 0.2 kg, 0.1 kg and 0.3 kg respectively. A is moving with speed 2 ms^{-1} when it collides with B which is stationary. At the instant that A and B collide, C is at the point 0.8 m from the point of collision, and moving towards A and B with speed 1.7 ms^{-1}. Immediately after the collision B starts to move with speed 2.4 ms^{-1} (see diagram).

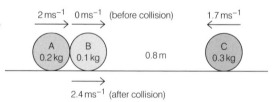

i) Find the speed of A immediately after the collision, and state its direction of motion. [4]

Spheres B and C collide Ts after the collision between A and B.

ii) Find the value of T. [6]

iii) Show that the speeds of B and C immediately before they collide are 2.3 ms^{-1} and 1.6 ms^{-1} respectively. [2]

Immediately after the collision between B and C, the particles move away from each other with speeds v ms^{-1} and v/28 ms^{-1} respectively.

iv) Find the value of v. [4]

CHAPTER 2
Exercise 3

1 **a)** $\rightarrow$ 5.62 $\uparrow$ 13.91
　 b) $\rightarrow$ 6.85 $\downarrow$ 2.22
　 c) $\leftarrow$ 10.23 $\uparrow$ 15.17
　 d) $\leftarrow$ 16.54 $\downarrow$ 15.98

2 **a)** $\begin{pmatrix} 18.74 \\ 20.81 \end{pmatrix}$　 **b)** $\begin{pmatrix} -107.86 \\ 52.60 \end{pmatrix}$

3 **a)** $\rightarrow$ 5.64 $\uparrow$ 2.05
　 b) $\rightarrow$ 14.86 $\uparrow$ 13.38
　 c) $\rightarrow$ 19.54 $\uparrow$ 46.03

4 **a)** 8.45   18.13
　 b) 11.41   3.71
　 c) 8   0
　 d) 25.22   19.70

5 **a)** 12.5   21.65 N
　 b) 8.32   39.13 N
　 c) W $\sin \theta$ W $\cos \theta$

6 29.54 N   5.21 N

7 **a)** 11.45   51.0°
　 b) 41.7   153.1°
　 c) 15.18   35.9° below positive
　　　　　　　 x-direction

CHAPTER 2
Exercise 4

1 307.7   66.8°

2 **a)** 7.56   0.996   **b)** −1.29   17.66
　 c) 0   0   **d)** 0   2P $\sin \theta$ − P

3 R = 13.7 N
　 at 39.1° to AB

4 P = 181.2 N   573.5

5 7.32 N   0

6 10.5   30.3

7 R = 292.1 N
　 θ = 5.9° to middle rope

8 11.95P
　 at 56.5° to AB

9 R = 4
　 P = 2($\sqrt{3}$ − 1) = 1.464

10 R = 18.4 N

CHAPTER 2
Revision Exercise

1 **a)** P = 6.189
　 b) 9.317

2 16.95   11.35

3 $\rightarrow$ 2.5 $\uparrow$ 4.33
　 Magnitude = 5   direction = 60°

4

5 14.42 N   56.3°

6 Magnitude = 11.36
　 at 22.4° to x direction

7 P = 20.92   17.14

8 149.1 N   16.83

9 P $\sin 20°$ + Q $\sin 50°$ = 16 $\sin 45°$
　 P = 16.96   Q = 7.20

10 R = 16.81
　 at 74.2° to 8 N force

CHAPTER 3
Exercise 1

1 **a)** Not in equilibrium
　 b) Equilibrium
　 c) Equilibrium

2 Q = 5$\sqrt{2}$ P = 5

3 θ = 41.4°   P = 5.29
　 A force of 5.29 N acting vertically down

4 20 N   17.3 N

5 26.69 N   9.34 N

6 1.173 N   0.353 N

7 5 N   8.66 N

8 3470 N

9 97.86 N   20.57 N

10 **b)** 24 N   10 N

11 **a)** θ = 30°   **b)** θ = 18.4°

CHAPTER 3

Exercise 2

1. a) 9500 N   b) 9000 N   c) 4500 N

2. a) 28 N   b) 22 N   c) 35 N   d) 6 N

3. a) 35 N   b) 22.5 N

4. a) i) 77.27 N   ii) 20.71 N
   b) i) 57.96 N   ii) 36.23 N

5. a) 72.09 N
   b) i) 489.07 N   ii) 176.04 N
   c) 278.42 N

6. i) 80 N   ii) 14.83 N   iii) 40 N

CHAPTER 3

Exercise 3

1. 0.577

2. 74.8 N

3. a) 0.364   c) 29.32

4. a) $F = 1$ (equilibrium)
   b) $F = 4$ (limiting equilibrium)
   c) $F = 4$ motion

5. i) 30 N   ii) 17.82 N
   iii) 20.92   iv) 1.17

6. b) $R = 34.49$ N $F = 51.75$ N
   c) $T = 57.10$ N
   d) block lifts off table

7. a) 5.72 N   b) 19.34 N

8. 1009.1 N
   $T_{min} = 970.14$ $\theta = 14.04°$

CHAPTER 3

Revision Exercise

1. $P = 13.3$ N $Q = 45.58$ N

2. 1000 N   866 N

3. a) 20 N   b) 14.82 N   c) 0.767

4. i) 5 N   36.9°
   ii) 4 N horizontally to right

5. a) 1026 N   b) 0.364

6. a) 19.38 N   b) 23.30 N

7. a) 2500 N   b) 22 500 N

8. iii) a) 0.75 N   b) 0.13 N

9. a) 9.71 N   b) 27.05 N   c) 0.359

10. a) 36.94 N   b) 0.781
    c) i) equilibrium $F = 23.06$
       ii) no equilibrium $F = 36.94$

11. 6.29 N

12. $T_{AB} = \dfrac{5}{2}w$ $T_{BC} = 4w$ $k = 4$

CHAPTER 4

Exercise 1

1. a) 3600 m   b) 40 m/min

2. a) 0.235 m/s   b) −0.141 m/s
   ii) constant speed during motion;
   stops exactly at desired point

3. b) 1425 m
   c) uniform acceleration and braking

4. a) 7.5 m/s   c) 0.9375 m/s^2
   d) 270 m

5. a) Pizza shop $\rightarrow$ A $\rightarrow$ Pizza
   shop $\rightarrow$ C $\rightarrow$ B $\rightarrow$ Pizza shop

6. a) 12 m/s due W
   b) 7.64 m/s due N
   c) 12 m/s due S
   d) 10.8 m/s in SW direction
   e) 12 m/s due E
   f) 0

7. a) 240 km/hr due E
   b) 240 km/hr due S
   c) 240 km/hr on bearing 323°
   134 km/hr on bearing 141°

CHAPTER 4

Exercise 2

1. 17 m/s   75 m

2. a) 4   0   −4   c) 60   64   60

3. a) 19.2   b) 28.8   c) 9.6 m

4. $u = 8$, $a = -1.6$

5. a) $t = 12$ s   b) $t = 24$ s   c) 216 m

6. 32.5 m

7 3 m/s^2 2s

8 19.35 s   599.2 m

9 33.25 m

10 a) $u = 15$, $a = -0.4$ **b)** 281.25 m

11 Assume uniform acceleration
0.149 m/s^2 27.9 s

12 Uniform rate of braking
−6.58 m/s^2
70.9 m

13 35 s   15 m/s

CHAPTER 4
Exercise 3

1 39.6 m/s

2 11.5 m

3 d) 4.95 s

4 5.099 s   35.97 m/s

5 41.6 m
No air resistance

6 0.0256 s

CHAPTER 4
Revision Exercise

1 a) 5 m/s   **b)** 10 s

2 i) 1 m   **ii)** 0.6 m

3 Assuming constant acceleration
3.47 m/s^2

4 b) 10.549 s   188.7 m

5 i) 16 m   **ii)** −2 m/s^2 **iii)** 14 m/s

6 b) 2975 m
c) constant acceleration

7 a) $u = 15$; $a = -0.4$
b) 37.5 s   281.25 m

8 a) 4.764 s   **b)** 30.7 m/s
c) no air resistance

9 15 s

10 a) 40 m   **b)** 14 m/s
c) $\frac{10}{7}$ s   **d)** $\frac{20}{7}$ s

11 a) i) $4t + t^2$ **ii)** $2t + 1.5t^2$ **b)** 4 s
c) 12 m/s in direction $\overrightarrow{AB}$
14 m/s in direction $\overrightarrow{BA}$

12 i) 270 m   **ii)** 30 m   60 m
iii) 0.6 m/s   **iv)** 1.4 m/s

CHAPTER 5
Exercise 1

1 a) no acceleration   **b)** acceleration

2 12 N

3 a) 1.5 m/s^2 **b)** 6N

4 a) 6 N   **b)** 9.50 N

5 75.36

6 $\mu = \frac{8}{49}$

7 29.7 N

8 0.551

9 3.98 m/s^2

10 b) 20 N   **c)** 49   **d)** $\frac{20}{49}$
e) 36.32   **f)** 2.47 m/s^2

11 a) 2.03 m/s^2 **b)** 62.44 N

12 a) 4×10^{14} **b)** 3080 N

13 660 N

14 2 m/s^2 down

15 a) 3051 N   **b)** 791 N

16 1937.5 N

17 0.131
No air resistance; constant value of μ on
slope giving constant acceleration

CHAPTER 5
Exercise 2

1 b) 14   **c)** 11 520

2 b) $28 - T = 5a$ $T = 2a$
c) $a = 4$ $T = 8$
Table probably not smooth

3 a) 1600 N   **b)** 370 N

4 0.6 m/s^2 23.92 N

5 $T = 13.65$ N, $m = 1.3$ kg

6 0.28 m/s²   2.39 s
Light inextensible string; smooth peg; no
air resistance

7 1.4 m/s²   6.72 N

8 **a)** 16.6 N
b) 4.6 N   0.0587
Light inextensible string; smooth pulley

9 3.92 m/s²   12.348 N

10 0.406 m/s²   20.41 N

11 **a)** $\frac{273}{40}\sqrt{2}$ N   **b)** 0.173 m/s²

12 **a)** 0.7 m/s²   **b)** 68.25 N   **c)** 63.7 N

13 **a)** 19.72 N   **b)** 40.19 N

14 **a)** $\frac{54}{10}$ mg   **b)** $\frac{43}{10}$ mg   **c)** $\frac{4}{3}$

    d) Light inextensible strings; smooth pulleys

CHAPTER 5
Revision Exercise

1 **b)** 39.2 N   **c)** 12 N   **d)** $\frac{15}{49}$

2 1600 N

3 **a)** 192 N   **b)** $\frac{17}{15}$ m/s² down

4 **a)** −4.409 m/s²   **b)** 88 200 N
Constant acceleration; ignoring
    friction/air resistance

5 **a)** $\frac{8}{5}$ mg   **b)** $\frac{3}{5}$ mg
Smooth pulley, inextensible string

6 **a)** 166.94 N   **b)** 580.2 N

7 **a)** 1.249 m/s²   29.21 N
b) 0.498 m/s²   28.43 N

8 **a)** $D = \dfrac{45\,000}{v}$ **b)** $R = 0.7v^2$
c) 0.681 m/s²   **d)** 40.06 m/s

9 **b)** 4.56 m/s²   **c)** 62.13 N
no friction
friction would reduce acceleration;
not affect normal

10 **a)** 0.854 m/s²   **b)** 3.45 s
c) 2.92 m/s

11 **a)** 4.9 m/s²   2.94 N
b) 4.43 m/s²   **c)** 3.08 m
d) 1.103 s

12 **a)** $4g - T = 4a$
b) $T - 3g = 3a$ 33.6
c) 1.296 m/s
light inextensible string; smooth pulley

13 **a)** 87 750 m/s²   **b)** 0.000513 s
c) 5265 N
constant retardation

14 **a)** 0.934 m/s²
b) Smooth surface/skier = particle
c) g varies over Earth's surface

15 **b)** R = 588 N   F = 24 N   T = 30 N

16 **a)** 1.5 m/s²   **b)** 3 m/s
c) 33.2 N   **d)** 24.2 N
e) Light inextensible string, smooth
    pulley, no air resistance
Acceleration of system ≈ $\frac{1}{6}$ acceleration of
Earth.
Motion takes ≈ $\sqrt{6}$ times longer.

CHAPTER 6
Exercise 1

1 **a)** 5 kg   **b)** 250 m/s   **c)** 1.2×10^6 Ns

2 5.94 Ns

3 4.4 m/s

4 4.88 m/s

5 **a)** 6.25 m/s²   **b)** 4.98 m

6 $\frac{1}{3}$ m/s   3 m/s

7 **a)** $2u$ in reverse direction to before
    collision
b) $\frac{8}{3}u$

8 12 kg

9 12 000 kg   18 000 kg

10 **a)** 10 m/s   **b)** 41.67 m/s²
c) 4.25

11 **a)** 9.899 m/s   **b)** 0.89995 m/s
c) 2.025 m/s²   **d)** 13 000 N

12 **a)** $2u$ **b)** $\frac{9}{10}u$ **c)** $\frac{7a}{6u}$

CHAPTER 6
Revision Exercise

1 3.75; 4.75

2 $\frac{13}{10}u$ $\frac{19}{10}u$

3 $\frac{1}{3}u$

4 800

5 3.85 m/s

6 i) $3u$ ii) $\frac{10}{3}$

7 a) 3.3875 b) 5.55 m/s up
2.45 m/s down
c) $\frac{13}{60}$ up d) 1.816 s

8 a) 3 m/s b) 160 kg
Treating skaters as particles;
constant speed between collisions

9 i) 2.5 m/s ii) 8 iii) 4.8 s

CHAPTER 7
Exercise 1

1 a) 40 m b) 44 m/s c) 34 m/s^2

2 a) −7.5 m/s −144 m/s
b) $t = 2$ 8 m

3 a) 4 b) 0.75 m/s^2

4 a) $v = 0.45t^2 - 2.4t + 2.4$
$a = 0.9t - 2.4$
b) $t = \frac{4}{3}, 4$ $s = \frac{64}{45}$, 0 respectively
c) 0.719 m/s 0.15 m/s

5 b) 0 $2\alpha + 72\beta = 0$
c) $\alpha = \frac{3}{4}$ $\beta = -\frac{1}{48}$ 72 m

6 a) $v = 1.2t^3 - 2.4t$ $a = 3.6t^2 - 2.4$
b) 20.16 Ns c) 24 N

7 b) $x = 8.4, v = 3.9$ $x = 16.6, v = 3.9$

CHAPTER 7
Exercise 2

1 a) 0.36 m/s b) 0.54 m

2 a) $v = 0.6 - 0.0006t^3$
b) $x = 2 + 0.6t - 0.00015t^4$
c) 8.5 m

3 $v(t) = t^3 + 5t + 8$
$x(t) = \frac{1}{4}t^4 + \frac{5}{2}t^2 + 8t$
30
66.75 36.75 m

4 a) $t = \frac{1}{3}$, 5
c) 8 m/s^2
d) $s(t) = t^3 - 8t^2 + 5t + 2$

5 a) $v(t) = \frac{1}{3}\alpha t^3 + \beta t + 0.4$
$5(t) = \frac{1}{12}\alpha t^4 + \frac{1}{2}\beta t^2 + 0.4t$
b) $\alpha = 3.6; \beta = -1$
c) 29.8 m/s $8\frac{1}{3}$ m/s

6 a) $a(t) = 12 - 6t$
b) $x(t) = 6t^2 - t^3$ $t = 6$

7 $t = 7$ s −0.2 m/s 29.9 m/s

8 b) $v = 5t - t^2$ $t = 0$ or 5
c) $20\frac{5}{6}$

9 a) $v_A = 0.3t^2 + 0.4t$
b) $v_B = 0.3t^2 + 0.3$
$x_B = 0.1t^3 + 0.3t + 2$
c) $t = 4$ s 9.6 m d) 5.62 m/s

CHAPTER 7
Revision Exercise

1 b) −4 m/s^2

2 i) $a = 3.2 - 0.15t$
$x = 1.6t^2 - 0.025t^3$
ii) 89.6 m

3 a) $v = 0.6t^2 - 4.8t + 7.2$
$a = 1.2t - 4.8$
b) $t = 2$ or 6
d) 0.8 m/s e) 2.4 m/s

4 i) $v(t) = 0.48t - 0.012t^2$
$x(t) = 0.24t^2 - 0.004t^3$

5 i) $a = -0.6t^2 - 1.8t - 6$
$x = 0.05t^4 - 0.3t^3 - 3t^2$
ii) −81.25 m

6 i) $v = 6t^2 - 4t^3$ $a = 12t - 12t^2$
ii) $t = 2$ iii) 8 m/s
iv) $t = 1.5$ v) 2 m/s

7 b) 125 m   **c)** 27 m

8 a) $x_P = 2t + 0.002t^3$ **b)** $x_Q = 5.75(t - 10)$
   d) Q is 1 m ahead of P

9 a) $v = \frac{-1}{20}t^2 + V$ **b)** $V = 5$ **c)** $33\frac{1}{3}$ m

10 a) $v_A = 0.03t^2 + 0.05t + 0.04$
      $v_B = 0.04t^2 + 0.02t$
   b) B is closer to O

REVISION 1
Forces

1 65.74 N at 92.4° to x direction

2 4.71 N   24.2°

3 P = 181.2 N   573.5 N

4 13.73 N at 39.1° to AB

5 $\theta = 30°$ 2.80 N

6 a) i) 5.321   **ii)** 6.907
   b) 8.72 N   **c)** 52.4°

7 angle between P and Q = 63.2°
   7.83 N

REVISION 2
Equilibrium

1 31.1 N

2 3.756 N   *not* limiting

3 T = 391.6 N   R = 251.8 N

4 Q = 4.257 N   P = 1.456 N
   horizontally to the left

5 $m = 30$ kg   T = 294 N

6 R = 14 199 N   F = 1805 N
   0.268

7 15.63N   15.07 N

9 T = 2.573 N   0.257 N

11 $\theta = 25.66°$ T = 21.71 N

12 i) 219.6 N   **ii)** 74.98 kg

13 iii) 0.346

14 0.703

15 0.546 kg

REVISION 3
Kinematics

1 2.77 m/s

2 29.1 m

3 $T = 2$ s $d = 60$ m

4 a) −0.096 Ns   **b)** −0.024 N

5 6.53 s

6 i) 40 s   **ii)** 24.85 m/s

7 i) 3 m/s   **ii)** $-\frac{2}{3}$ m/s
   Hit another ball or cushion

8 a) $v = 9 - 0.03t^2$
   b) $x = 9t - 0.01t^3$
   c) −18 m/s
   d) $60\sqrt{3}$ m

9 $u = 39.2$ m/s
   i) Constant acceleration so gradient
      should be −9.8
   ii) All are wrong

10 i) 40
   ii) 0.0667 m/s²
   iii) 35 m

11 a) 10 s   **b)** 200 m

12 i) 1.607 s
   ii) v_A 5.3 m/s up
      v_B 14.7 down

13 6.7945 s   65.9 m

14 54.72 m

15 a) 3.125 m/s²
   b) 6.25 m/s²
   c) 22.5 m
   d) constant acceleration – unlikely
      because of gears
      Allow for different accelerations in
      different gears

16 4.91 s   48.66 m from A

17 a) $x(3) = 3u + 4.5a$
$x(4) = 4u + 8a$

c) $37 = u + \frac{15}{2}a$

d) $u = 7; a = 4$

18 a) $u = 10$　　$a = 0.2$　　**b)** 250 m

19 1.2 m/s^2　　3.75 m　　0.75 m

20 0.8 m/s^2　　317.5 m

21 $a = 0.5$ m/s^2
accelerating time = 20 s
10 m/s

22 a) 0.267 m/s^2　**b)** $t = 60$　**c)** 480 m

REVISION 4
Newton's Laws of Motion

1 1640 N

2 i) 3136 N　　**ii)** 3264 N

3 36.75 N

4 0.297

6 1.6 N　　$\frac{8}{45}$ kg

7 3.11 m

8 19.75 s

9 i) 200 N　　**ii)** 12.5 N

10 6.86 kg　　8.73 m

11 ii) 96.9 kg

12 a) i) 100　　**ii)** 400　　**b)** 0.22 m/s^2

13 20.5 N

14 2.70 m/s　　587.1 N

15 i) 1.778 m/s^2　　**iii)** 567.1 N

16 18 N　　0.265　　0.106 m

17 b) $8mg - T = 8ma$
$S - 3mg = 3ma$
$T - S = 5ma$

c) $\frac{11}{2}mg$　　$\frac{63}{16}mg$

d) Light smooth pulleys; light
inextensible strings

18 $a = \frac{1}{10\,000}(X - 600)$
$X < 600$　　$X < 60$

19 a) iii) 1.96 m/s^2　　**iv)** 3.429 m/s
b) i) -4.312 m/s^2
c) $\mu = \frac{1}{4}$

20 1.320 N　　0.5399 N　　40°

REVISION 5
Momentum

1 $v = 0.4$ m/s

2 a) $\frac{5}{2}u$　　**c)** $\frac{1}{5}T$

3 $m = 0.1$　　$m = \frac{24}{65}$

4 1.7 m/s and 2.2 m/s

5 $m = 4$ kg

6 a) 0.6 m/s　　**c)** 3.3 m long

7 a) 8 s　　**b)** 12 m/s　　24 m/s
c) 16 m/s

8 $8m$ and $2m$

Sample exam paper

1 i) 0.5 ms^{-2}　　**ii)** 2.92

2 i)

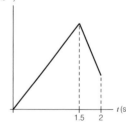

ii) 4.7 ms^{-1}　　**iii)** 4.85 m

3 i) 3　　**ii)** 2.14 N

4 i) 1.21　　**ii)** 10.2

5 ii) 1.4, 1.8　　**iii)** 2.8 ms^{-2}

6 i) 4.8 ms^{-1}　　**ii)** 6.4 ms^{-1}
iii) 1.33 s　　**iv)** 6.67 s

7 i) 0.8 ms^{-1} in the same direction as
before
ii) 0.2　　**iv)** 2.8

INDEX